CLIMATE CHANGE

WHAT EVERYONE NEEDS TO KNOW®

JOSEPH ROMM

OXFORD
UNIVERSITY PRESS

OXFORD
UNIVERSITY PRESS

Oxford University Press is a department of the University of
Oxford. It furthers the University's objective of excellence in research,
scholarship, and education by publishing worldwide.

Oxford   New York

Auckland   Cape Town   Dar es Salaam   Hong Kong   Karachi
Kuala Lumpur   Madrid   Melbourne   Mexico City   Nairobi
New Delhi   Shanghai   Taipei   Toronto

With offices in

Argentina   Austria   Brazil   Chile   Czech Republic   France   Greece
Guatemala   Hungary   Italy   Japan   Poland   Portugal   Singapore
South Korea   Switzerland   Thailand   Turkey   Ukraine   Vietnam

Oxford is a registered trademark of Oxford University Press
in the UK and certain other countries.

Published in the United States of America by
Oxford University Press
198 Madison Avenue, New York, NY 10016

Library of Congress Cataloging-in-Publication Data
Romm, Joseph.
Climate change : what everyone needs to know / Joseph Romm.
pages cm.—(What everyone needs to know)
Includes index.
ISBN 978–0–19–025017–1 (paperback)
1. Climatic changes—Health aspects.   2. Climatic changes—Economic
aspects.   3. Human beings—Effect of climate on.   4. Global
warming.   I. Title.
RA793.R66 2015
363.738'74—dc23
2015020730

5 7 9 8 6
Printed in Canada
on acid-free paper

To Antonia

CLIMATE CHANGE

WHAT EVERYONE NEEDS TO KNOW®

CONTENTS

2 Extreme Weather and Climate Change 31

3 Projected Climate Impacts   73

PREFACE

Why you need to know about climate change

Climate change will have a bigger impact on your family and friends and all of humanity than the Internet has had. Imagine if you knew a quarter-century ago how information technology and the Internet were going to revolutionize so many aspects of life. Imagine how valuable that knowledge would have been to you and your family. It turns out that we have such advanced knowledge of how climate change will play out over the next quarter-century and beyond. The purpose of this book is to provide you that knowledge.

Climate change is now an existential issue for humanity. Serious climate impacts have already been observed on every continent. Far more dangerous climate impacts are inevitable without much stronger action than the world is currently pursuing, as several major 2014 scientific reports concluded.

Since everyone's family will be affected by climate change—indeed, they already are—everyone needs to know the basics about it, regardless of their politics. Many of the major decisions that you, your family, and friends will have to make in the coming years and decades will be affected by human-caused climate change. Should you own coastal property? Should you plan on retiring in South Florida or the Southwest or the Mediterranean? What occupations and career paths make the most sense in a globally warmed world, and what should students study? Should climate change affect how you invest for the future?

This book will explore those questions as well as the more basic ones everyone needs to know the answers to, such as why are climate scientists so confident that humans are the primary cause of recent warning? Which extreme weather events are being made worse by climate change and which are not? What are the core energy-related climate solutions? Since climate is always changing, why worry about what is happening now? What actions are the major polluters—China, the United States, and the European Union—taking to reduce emissions of heat-trapping greenhouse gases?

Also, this is the first book to examine one of the most important climate questions of all: "Does carbon dioxide at exposure levels expected this century have any direct impacts on human health or cognition?" You might think that the answer to this obvious question would be well researched by now, especially since a great many people are routinely exposed to such levels today *indoors*. But it is only very recently that scientists have done the relevant studies. Even more surprising, preliminary research—including studies by Lawrence Berkeley National Laboratory and the Harvard School of Public health—suggests that the answer is "yes"! And that has huge implications for you and your family now.

Under different circumstances, you might not have needed to become especially knowledgeable about climate change. Consider the case of the Earth's ozone layer, which protects us from dangerous ultraviolet light. In 1974, climate scientists figured out that chlorofluorocarbons (CFCs) were destroying the ozone layer. Americans and Scandinavian countries voluntarily banned CFC use in spray cans within 5 years. A few years after that, President Ronald Reagan, Vice President George H.W. Bush, and British Prime Minister Margaret Thatcher played an instrumental role in bringing about an international treaty banning CFCs. Decades later, the ozone layer has still been preserved, and you do not need to think about it at all.

However, climate change action has not followed that rational trajectory. Scientists have known for over a century that human-caused greenhouse gases would warm the planet. Four decades ago, climate scientists began seriously sounding the alarm about the dangers posed by unrestricted emissions of greenhouse gases such as carbon dioxide from the burning of fossil fuels. In 1977, the U.S. National Academy of Sciences warned that unrestricted greenhouse gas emissions might raise global temperatures as much as 10°F [5.5°C] and raise sea level 20 feet. The Academy, the nation's most prestigious scientific body, was chartered by Abraham Lincoln to give advice to the nation on scientific matters.

In 1988, the nations of the world came together to task the top scientists of the world with regularly summarizing and reporting on the latest research and observations. The central purpose of the resulting United Nations Intergovernmental Panel on Climate Change (IPCC) was to provide the best science to policymakers. In the ensuing years, the science has gotten stronger, in large part because observations around the world confirmed the vast majority of the early predictions made by climate scientists.

At the same time, many cornerstone elements of our climate began changing far faster than most scientists had projected. The Arctic began losing sea ice several decades ahead of every single climate model used by the IPCC, which in turn means the Arctic region warmed up even faster than scientists expected. At the same time, the great ice sheets of Greenland and Antarctica, which contain enough water to raise sea levels ultimately 25–80 meters (80–260 feet), have begun disintegrating "a century ahead of schedule," as Richard Alley, a leading climatologist put it in 2005. In 2014 and 2015, we learned that both ice sheets are far less stable than we realized, and they are dangerously close to tipping points that would lead to irreversible collapse and dramatic rates of sea-level rise.

In the last several years, we have experienced a spate of off-the-charts extreme weather events that scientists had predicted decades ago—heat waves, droughts, wildfires, superstorms, and super storm surges. Meteorologist and former hurricane hunter Dr. Jeff Masters said in 2012, "This is not the atmosphere I grew up with." There is an ever-expanding body of scientific literature that clearly shows that greenhouse gases are fundamentally altering the climate and sharply boosting the chances for many types of extreme weather events.

For these reasons, the IPCC, the U.S. National Academy, the United Kingdom's Royal Society, as well as many other scientific and international organizations have released increasingly urgent warnings of the dangers of inaction as well as stronger and stronger calls to action. Although this has all helped restart a global conversation on climate change, human-caused emissions of greenhouse gases have continued their rise almost unabated. In fact, growth in emissions has accelerated since the year 2000.

Because global temperature rise and other impacts are driven by cumulative emissions of greenhouse gases, those soaring emissions have kept the world on track for the worst-case scenarios of climate change. The slow pace of national and global action in response to the scientific reports written for policymakers has led to more scientists communicating directly to the public. In 2010, Ohio State scientist Lonnie Thompson explained why climatologists had begun speaking out more: "Virtually all of us are now convinced that global warming poses a clear and present danger to civilization."[1]

The good news is that, finally, in November 2014, the world's largest polluter, China, announced a deal with the United States whereby China would peak in carbon dioxide emissions by 2030. In that deal, the United States committed to cut greenhouse gases 26%–28% below 2005 levels by 2025. A few weeks earlier, the entire European Union had pledged to cut total emissions 40% below 1990 levels by 2030. So, by the end of 2014, countries representing more than half of all emissions worldwide had made serious commitments to limit global warming pollution. Many other countries have made commitments in 2015, including Japan, Russia, and Mexico.

This deal does not mean dangerous climate change will be avoided. Nevertheless, it is a game-changer in the sense that before this deal, neither the United States nor China was seriously in the game of trying to stave off climate catastrophe. However, now both countries are. In addition, the United States-China deal, combined with the commitments by the European Union and other countries, greatly increases the chance of a global agreement in Paris, December 2015, one that would finally shift the world off of an emissions path headed toward an unimaginable 6°C (11°F) total warming and on to an emissions path that would keep total warming below the catastrophic level of 4°C (7°F).

Because action has been so delayed for so long, however, humanity cannot avoid very serious climate impacts in the coming decades—impacts that will affect you and your children. Therefore, you need to understand what is coming so you and your family will be prepared. It is entirely possible, if not likely, that climate change will transform the lives of your children more than the Internet has. In some sense, the defining story of the 21st century is a race between the impacts our cumulative carbon emissions will increasingly have on our climate system and humanity's belated but accelerating efforts to replace fossil fuels with carbon-free energy.

Will we stay on our current path and trigger amplifying feedbacks that cause further warming, pushing us closer to irreversible tipping points. Or will we instead act quickly enough to avoid the very worst impacts? Leading scientists and governments say that would mean keeping total warming as close to 2°C (3.6°F) as possible and preferably below it. Meeting such a warming target would require all nations to replace fossil fuels with clean energy at an even faster rate than we are currently planning—and for total global carbon dioxide emissions to be zero (or negative) by century's end.

The story of the ongoing clean energy revolution is thus an inextricable part of the climate change story. For instance, in the past quarter century, the price of solar-powered electricity from photovoltaic panels has dropped by 99%, which has been accompanied by an equally impressive 60% annual increase in global solar capacity. This trend is certain to continue, in part because in its 2014 climate deal with the United States, China also committed to more than double its share of carbon-free sources of energy (such as solar power, wind power, nuclear power, and hydropower) by 2030. In 2014, the European Union similarly adopted a binding target to increase its share of renewable energy to at least 27% of its 2030 energy consumption. That is approximately double its current renewable energy share.

These commitments mean that the recent explosive growth and price drops experienced by renewable energy sources such as solar and wind will continue for decades to come. In addition, it means the long-predicted ascendance of carbon-free energy has now begun in earnest. Indeed, just 1 week after its pledge to peak carbon dioxide emissions in 2030, China announced that its peak in coal consumption would be in 2020. This is a complete reversal of Chinese energy policy, which for two decades has been centered on building a coal plant or more each week. Now, however, they will be building the equivalent in carbon-free power every week for decades, while the construction rate of new coal plants grinds to a halt over the next several years. My June 2015 trip to meet with leading Chinese climate and energy experts made clear the country is likely to beat its stated targets—with carbon dioxide peaking by 2025 and coal use peaking as early as right now.

The accelerating global shift away from carbon-intensive forms of energy, agriculture, and transportation will have effects on you and your family (and everyone else) almost as great as the impacts from the climate change itself. It will affect the cars we drive, the products we buy, the homes in which we live, the industries that succeed and fail in the marketplace, the global financial system, and ultimately even the foods we eat.

A key goal of this book is to save you time. There is far too much information on climate science, clean energy solutions, and global warming politics for you to stay current. Now, however, everyone needs to follow this issue and have an informed opinion in order to participate in the growing conversation on the most important issue of the decade and the century. More importantly, everyone needs to understand how climate change—and our response to it—is going to directly affect their lives and the lives of their family in the years and decades to come.

This book will not enter into the unproductive political debate over the science. Rather, it takes as a starting point the overwhelming consensus of our top global experts and governments, as laid out in the recent Intergovernmental Panel on Climate Change summary reviews of the literature, culminating with the November 2014 "Synthesis Report." The 2014 Report issued their bluntest statement yet to the world: Cut carbon pollution sharply starting now (at a very low cost) or risk "severe, pervasive and irreversible impacts for people and ecosystems." In addition, this is a particularly apt time for such a book because in 2014 we saw an unusual number of highly credible—and uncharacteristically blunt—major reports, including ones by the American Association for the Advancement of Science, the U.S. National Academy of Sciences, and the United Kingdom Royal Society. These reports, and the Congressionally mandated U.S. National Climate Assessment, provide a solid basis for explaining the physical science behind global warming, the projected impacts we are facing, and how to avoid catastrophe.

This book is aimed at general readers in North America, Europe, and around the world who are interested in understanding what climate change means for them and their families, as well as those interested in joining the growing public debate and discussion. The book will cover our current understanding of climate science as well as the most salient projected impacts that we are facing. It will also examine climate solutions, especially in the crucial energy sector. Finally, it will examine the political and policy issues surrounding climate change.

I have been deeply involved in climate science, solutions, and policy for a quarter century, and I have focused on effectively communicating aspects of climate change to a general audience for a decade. I first became interested in global warming in the mid-1980s while studying for my physics Ph.D. at the Massachusetts Institute of Technology. I researched my Ph.D. thesis on the physical oceanography of the Greenland Sea at the Scripps Institution of Oceanography and had the chance to work with Dr. Walter Munk, one of the world's top ocean scientists. The impact of climate change on the oceans was already a concern at Scripps in the 1980s.

In the mid-1990s, I served for 5 years in the U.S. Department of Energy. As acting assistant secretary for energy efficiency and renewable energy, I oversaw what was then the world's largest portfolio of research and development, demonstration, and deployment of low-carbon technology—$1 billion aimed at advanced energy efficiency technologies in buildings, industry, and transportation as well as every form of renewable energy. I helped develop a climate technology strategy for the nation. While working with leading scientists and engineers at our national laboratories, I came to understand that the technology for reducing our emissions was already at hand and at a far lower cost than was widely understood. After the Department of Energy, I worked with some of the nation's leading corporations, helping them to make greenhouse gas reductions and commitment plans that reduced emissions by millions of tons while also boosting their profits.

After my brother lost his Mississippi home in the August 2005 Hurricane Katrina storm surge, he asked me for advice on whether he should rebuild there. So I started interviewing climate experts and attending climate seminars, and I began to read the scientific literature. Dozens of our top climate scientists impressed upon me the fact that the climate situation was far more dire than I had realized, far more dire than 98% of opinion makers and politicians understood—a situation that remains true today.

This knowledge led me to shift from helping companies make greenhouse gas cuts to focusing on a better understanding of climate science and solutions, and I was able to concentrate on how to communicate this issue to policymakers and the public. I became founding editor of ClimateProgress. org, which *New York Times* columnist Tom Friedman called "the indispensable blog." Since 2006, I have written millions of words on every aspect of climate change, I have reviewed hundreds of studies and reports, and I have interviewed the vast majority of leading experts on climate change and solutions. In 2009, *Time* magazine named me a "Hero of the Environment" and "The Web's most influential climate-change blogger."

For the last few years, I have also had the privilege to be Chief Science Advisor for the TV series, "Years of Living Dangerously," the first climate change docu-series ever to appear on U.S. television. This has given me the chance to work with some of the best communicators in the country, from James Cameron to former "60 Minutes" producers. Season One of "Years of Living Dangerously," which aired on Showtime in mid-2014, won the prime time Emmy for "Outstanding Non-Fiction Series." *The UK Guardian* called the series, "Perhaps the most important climate change multimedia communication endeavor in history." Season Two is scheduled to air in 2016 on National Geographic Channel.

In the coming years, climate change will become a bigger and bigger part of all our lives. It is literally the story of the century, and, for better or worse, you and everyone you know will increasingly become a part of that story. Here is what you and your family need to know to navigate your future.

ACKNOWLEDGMENTS

This book would not have been possible without the opportunity I have had over the past nine years to publish blog posts containing millions of words on climate science, policy, and solutions. In addition, the rapid feedback that I received online from readers and other science bloggers helped to greatly improve my writing skills since 2006.

That's why I first thank John Podesta—for believing in the idea of ClimateProgress.org and making it possible to launch the website. I want to thank everyone at the Center for American Progress Action Fund (CAPAF) who supported ClimateProgress.org over the years as the website grew and expanded and ultimately merged with ThinkProgress.org, and I would like to especially thank Neera Tanden and Judd Legum from ThinkProgress. Many people now and in the past have given me invaluable advice and support since the blog was launched, including Kari Manlove, Sean Pool, Brad Johnson, Faiz Shakir, Nico Pitney, Stephen Lacey, Andrew Sherry, Dan Weiss, and a variety of interns. I thank the IT wizards at CAPAF who have provided unmatched technical and design support for the site, and I especially thank George Estrada, Matt Pusateri, Nick Levay, and Josh Hill.

I would also like to thank the producers of the TV docuseries "Years of Living Dangerously," particularly Joel Bach, James Cameron, David Gelber, and Maria Wilhelm. Working on that show over the past five years has given me exposure to some of the best and most imaginative communicators in the world, which has taught me a great deal about communicating climate science and solutions to a general audience.

Although the final judgments in each answer provided in this book are my own, I am exceedingly grateful to everyone who has shared their wisdom about climate science and solutions with me over the past few years: Richard Alley, Richard Betts, Robert Brulle, Ken Caldeira, Josep Canadell, Benjamin Cook, Bob Corell, Heidi Cullen, Aiguo Dai, Mik Dale, Andrew Dessler, Noah S. Diffenbaugh, James Elsner, Kerry Emanuel, Peter Fox-Penner, Jennifer Francis, Andrew Frank, Peter Gleick, Saul Griffith, Nicole Hernandez Hammer, James Hansen, Melanie Hart, David Hawkins, Katharine Hayhoe, Peter Höppe, Andrew Jones, Tina Kaarsberg, Thomas Karl, Dan Kammen, Greg Kats, David Keith, Henry Kelly, Felix Kramer, Kalee Kreider, David Lawrence, Michael Mann, Wieslaw Maslowski, Jeff Masters, Bill McKibben, Philip Mote, Jeff Nesbit, Michael Oppenheimer, Naomi Oreskes, Stu Ostro, Jonathan Overpeck, Stefan Rahmstorf, Eric Rignot, Dave Roberts, Arthur Rosenfeld, Doug Rotman, Gavin Schmidt, Mark Serreze, Robert Socolow, John Sterman, Lonnie Thompson, David Titley, Kevin Trenberth, Till Wagner, Harold Wanless, and David Yarnold.

I appreciate the time that Bill Fisk, Joseph Allen Gardner, and Pawel Wargocki spent with me discussing their work on the crucial question of whether elevated CO_2 levels impact human cognition. I am very grateful to Fisk and Gardner for reviewing my answer. Special thanks go to the following individuals who reviewed all or part of the book: John Abraham, John Cook, Jon Koomey, and Ryan Koronowski. Their comments greatly improved the final product.

I would like to thank my editor at Oxford University Press, Chad Zimmerman, for giving me the opportunity to write this book, helping me craft it, and for making the entire process run smoothly. I would like to thank him and the rest of the staff at OUP for their enthusiastic support of this book from the very beginning.

I owe a permanent debt to my parents for sharing with me their unmatched language and editing skills over the years— and for teaching me the discipline of meeting a deadline. I would not be a blogger today were it not for the guidance and wisdom of my mother, Ethel Grodzins Romm—a woman whose transcendental nature cannot be fully understood until you meet her and she changes your life for the better, as she has for so many people, including me.

Finally, words cannot describe my gratitude for the gift of my daughter, Antonia, who every day provides me with unlimited inspiration. At the age of three, she started saying "blah, blah, blah." When I asked her if she knew what "blah, blah, blah" meant, she paused and said, "It's when Daddy says something that doesn't matter." Bingo. Until I had a daughter, I never realized just how often I said things that don't matter. But with her help, I've learned to do that less often, and I hope that is reflected in this book. Ultimately, it is the Antonias around the world who we must all bear in mind when we contemplate the consequences of our action—and our inaction— on the climate.

1

CLIMATE SCIENCE BASICS

This chapter focuses on climate science as it applies to what science can tell us about the changes we have observed to date and what caused them. The goal is to answer the key questions that people ask about the science.

What is the greenhouse effect and how does it warm the Earth?

The greenhouse effect has made life as we know it possible. The basic physics has been understood for well over a century. The sun pours out intense amounts of radiation across the electromagnetic spectrum, including ultraviolet and infrared. The sun's peak intensity is in visible light. Of the solar energy hitting the top of the atmosphere, one third is reflected back into space—by the atmosphere itself and the Earth's surface (land, ocean, and ice). The rest is absorbed, mostly by the Earth, especially our oceans. This process heats up the planet. The Earth reradiates the energy it has absorbed mostly as heat in the form of infrared radiation.

Some naturally occurring atmospheric gases let visible light escape through into space while trapping certain types of infrared radiation. These greenhouse gases, including water, methane (CH_4), and carbon dioxide (CO_2), trap some of the reradiated heat, so they act as a partial blanket that helps keep the planet as much as 60°F warmer than it otherwise would be, which is ideal for us humans.

At the dawn of the Industrial Revolution 250 years ago, CO_2 levels in the atmosphere were approximately 280 parts per million (ppm). Since then, humankind has been pouring billions of tons of extra greenhouse gases into the atmosphere, causing more and more heat to be trapped. The main human-caused greenhouse gas is CO_2, and the rate of growth of human-caused CO_2 emissions has been accelerating. Emissions today are six times higher than they were in 1950. Moreover, CO_2 levels have now hit 400 parts per million.

As a result, the Earth has warmed 1.5°F (0.85°C) since 1900. Most of this warming, approximately 1°F, has occurred only since 1970.

Why are scientists so certain the climate system is warming?

The world's leading scientists and governments have stated flatly, "Warming of the climate system is unequivocal" and a "settled fact." They have such a high degree of certainty the climate is warming because of the vast and growing amount of evidence pointing to such a conclusion.

In 2007, the United Nations Intergovernmental Panel On Climate Change (IPCC)—a scientific body with hundreds of the world's top scientists and climate experts—released its Fourth Assessment Report, which summarized thousands of scientific studies and millions of observations. That summary was approved line-by-line by the governments representing the overwhelming majority of the Earth's population. They concluded that warming "is unequivocal, as is now evident from observations of increases in global average air and ocean temperatures, widespread melting of snow and ice, and rising global average sea level."

Because science is based on observations, it is always subject to revision. On the one hand, evidence might arise that weakens our confidence in a theory. On the other hand, as the U.S. National Academy of Sciences wrote in its 2010 report, *Advancing the Science of Climate Change*:

"Some scientific conclusions or theories have been so thoroughly examined and tested, and supported by so many independent observations and results, that their likelihood of subsequently being found to be wrong is vanishingly small. Such conclusions and theories are then regarded as settled facts. This is the case for the conclusions that the Earth system is warming and that much of this warming is very likely due to human activities."

Scientists view warming as a settled fact because so much evidence points to that conclusion. For instance, the 1980s were the warmest decade on record at the Earth's surface. That record was then topped by the 1990s. And again, the 2000s were the hottest year on record. The year 1998 was the hottest on record until 2005, and then 2010 topped 2005, and then 2014 became the hottest year on record. Now 2015 is on track to top 2014 and become the hottest year on record by far.

Not only has the ocean's surface temperature increased in the past several decades, but the ocean's heat content has also increased. In addition, because the ocean is warmer, more water has evaporated from it, which leads to higher levels of humidity: this result has also been observed. With more water vapor in the air, you would expect more intense rainfall events and deluges, and, indeed, scientists have observed the increased frequency of these events. This warming has also been detected in the activity of plant and animal life. Spring is coming sooner across the globe, as observed in earlier and earlier blooming of plants. Likewise, all sorts of plant and animal species are shifting or migrating toward the poles or toward higher altitudes. Because the average temperature is rising, scientists expected the duration, severity, and frequency of heat waves to increase in a great many regions. This has also been observed, along with many other types of warming-related extreme weather events, as we will see in Chapter Two.

How does global warming increase sea levels and what has been observed to date?

One of the most visible and dangerous impacts from global warming is sea level rise. Human-caused warming has raised ocean levels on average several inches since 1900. In addition, the rate of sea level rise since the early 1990s has been almost 0.3 centimeters (0.12 inches) a year, which is double what the average speed was during the prior eight decades. Some of the most important contributors to sea level rise are accelerating.[2]

As one 2014 study explained, there are five main contributors to warming-driven sea level rise:

1. Thermal expansion
2. Changes in groundwater storage
3. Glacier ice loss
4. Greenland ice loss
5. Antarctic ice loss

Thermal expansion raises sea levels because the ocean, like all water, expands as it warms up and thus takes up more space. Warming-driven expansion is responsible for approximately half of the sea level rise in the past hundred years. In addition, around the globe, large amounts of land-based water, especially groundwater (such as is found in underground aquifers), is pumped out for farming and drinking. Because more groundwater is extracted than returns to the ground, that water also ends up in the world's oceans, which contributes to sea level rise.

Melting mountain glaciers also contribute to sea level rise, because frozen water that was trapped on land flows to the sea. Globally, some 90% of glaciers are shrinking in size. The previously land-locked ice ends up in the oceans, which boosts sea level rise. The cumulative volume of global glaciers began to decrease sharply in the mid-1990s. This coincides with a more than doubling of the rate of sea level rise.

Greenland and Antarctica are both covered with two enormous ice sheets. The Greenland ice sheet is nearly 2 miles (3 kilometers) thick at its thickest point and extends over an area almost as large as Mexico. If it completely melts, Greenland, by itself, would raise sea levels more than 20 feet. In 2012, a team of international experts backed by NASA and the European Space Agency put together data from satellites and aircraft to produce "the most comprehensive and accurate assessment to date of ice sheet losses in Greenland and Australia." They found that the Greenland ice sheet saw "nearly a five-fold increase" in its melt rate between the mid-1990s and 2011. The year 2012 in particular saw unusually high spring and summer temperatures in Greenland. NASA reported that year, "According to satellite data, an estimated 97% of the ice sheet surface thawed at some point in mid-July." Scientists told ABC News they had never seen anything like this before. In the summer of 2012, the Jakobshavn Glacier, Greenland's largest, moved ice from land into the ocean at "more than 10.5 miles (17 kilometers) per year, or more than 150 feet (46 meters) per day," another study found. The researchers pointed out, "These appear to be the fastest flow rates recorded for any glacier or ice stream in Greenland or Antarctica." By 2014, researchers were able to map Greenland's ice sheets using the European Space Agency satellite CryoSat-2, which can measure the changing height of an ice sheet over time. They found that since 2009, Greenland had doubled its annual rate of ice loss, to some 375 cubic kilometers per year.

The Antarctic ice sheet is vastly larger than Greenland—bigger than either the United States or Europe—and its average thickness is 1.2 miles (2 kilometers). The Antarctic ice sheet contains some 90% of all the Earth's ice. It would raise sea levels 200 feet if it completely melts. The West Antarctic ice sheet (WAIS) in particular has long been considered unstable because most of the ice sheet is grounded far below sea level—on bedrock as deep as 1.2 miles (two kilometers) underwater. The WAIS is melting from underneath. As it warms, the WAIS outlet glaciers become more unstable. In the future, rising sea levels themselves may lift the ice, thereby letting more warm water underneath it, which would lead to further bottom melting, more ice shelf disintegration, accelerated glacial flow, and further sea level rise, in an ongoing vicious cycle. A 2012 study found that Antarctica's rate of ice loss rose 50% in the decade of the 2000s. In 2014, researchers looked at measurements by the European Space Agency's CryoSat-2 satellite "to develop the first comprehensive assessment of Antarctic ice sheet elevation change." They concluded: "Three years of observations show that the Antarctic ice sheet is now losing 159 billion tonnes of ice each year—twice as much as when it was last surveyed." Two major studies from 2014 found that some WAIS glaciers have begun the process of irreversible collapse. One of the authors explains, "The fact that the retreat is happening simultaneously over a large sector suggests it was triggered by a common cause, such as an increase in the amount of ocean heat beneath the floating sections of the glaciers."

In late 2014, researchers reported the results of a comprehensive, 21-year analysis of the fastest-melting region of Antarctica, the Amundsen Sea Embayment. This region is approximately the size of Texas, and its glaciers are "the most significant Antarctic contributors to sea level rise." During those two decades, the total amount of ice loss "averaged 83 gigatons per year (91.5 billion U.S. tons)." This is equivalent to losing a Mount Everest's worth of ice (by weight) every 2 years. Coauthor Isabella Velicogna said, "The mass loss of these glaciers is increasing at an amazing rate."

Where does most of human-caused warming go?

The vast amount of overall human-caused warming—more than 90%—goes into heating the oceans, according to the latest climate science. Water has a tremendous capacity to store heat. The atmosphere stores only approximately 1% of man-made warming because it has a relatively poor heat storage capacity. The Earth's surface stores only approximately 2% of global warming because the land also cannot store heat the way the oceans do.[3]

Therefore, it is no surprise that in recent years, we have seen rapid warming in the oceans, according to several major studies. Those studies analyzed ocean temperature measurements from an array of global ocean buoys, bathythermographs (small underwater temperature probes), and other relevant data (such as sea level and surface temperatures). A 2013 study found a "sustained warming trend" since 1999 in the record of ocean heat content below 700 meters (2300 feet). The study found more total planetary warming in the past 15 years than the previous 15 years. The authors concluded that "recent warming rates of the waters below 700m appear to be unprecedented"—much higher than anytime in at least the last 50 years. That deep ocean warming has in turn been "contributing significantly to an acceleration of the warming trend." A 2014 study found that the upper 700 meters of the ocean have been warming up to 55% faster since 1970 than previously thought.

In 2015, U.S. National Oceanic and Atmospheric Administration (NOAA) released this chart of the change in global ocean heat content in the past six decades (Figure 1.1).

What fraction of recent global warming is due to human causes versus natural causes?

The latest science finds that *all* of the warming since 1970 is due to human causes. In September 2013, the United Nations Intergovernmental Panel on Climate Change (IPCC) released the first part of its Fifth Assessment Report, a summary report of the scientific literature. That summary was approved line-by-line by the governments representing the overwhelming majority of the Earth's population.

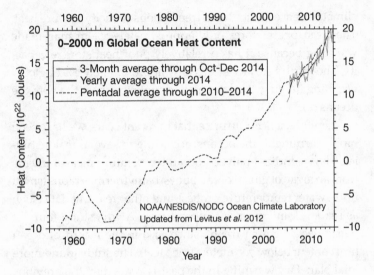

Figure 1.1 The oceans—where over 90% of global warming heat ends up—have warmed up very rapidly in recent decades. That warming shows no sign of slowing down in recent years.

Source: NOAA

The Panel concluded, "The best estimate of the human-induced contribution to warming is similar to the observed warming over this period" from 1951 to 2010. In other words, the best estimate is that humans are responsible for all of the warming we have experienced since 1950—based on a review of observations and analysis published in the scientific literature.

One reason the world's top scientists have confidence that humans are responsible for so much of the warming is that most of the naturally occurring things that affect global temperature would tend to be cooling the Earth. That is, in the absence of human activity and the warming that results from it, the planet would likely have cooled in recent decades. For instance, the sun's level of activity tends to have a modest, cyclical impact on global temperatures. In recent years, we have seen "the deepest solar minimum in nearly a century," as NASA explained in 2009—an unusually low level of solar activity that would otherwise be cooling the Earth slightly. Similarly, volcanic activity in recent decades has released particles that partially block the sun and also serve to cool the planet slightly. Finally, the underlying long-term trend for the Earth—driven largely by changes in our orbit—has been a very slow cooling. Human activity has overwhelmed all of these trends.

How certain are climate scientists that humans are the primary cause of recent warming?

Scientists are as certain that humans are responsible for most recent climate change as they are that cigarettes are harmful to human health. Studies reveal that some 97 of 100 actively publishing climate scientists agree with the overwhelming evidence that humans are causing global warming. The Climate Science Panel of the world's largest general scientific society, the American Association for the Advancement of Science, issued a report in March 2014, *What We Know*. That report described the scientific consensus this way:

> The science linking human activities to climate change is analogous to the science linking smoking to lung and cardiovascular diseases. Physicians, cardiovascular scientists, public health experts and others all agree smoking causes cancer. And this consensus among the health community has convinced most Americans that the health risks from smoking are real. A similar consensus now exists among climate scientists, a consensus that maintains climate change is happening, and human activity is the cause.

How do scientists know that recent climate change is primarily caused by human activities?

Scientists have such high confidence that human activity is the primary driver of recent warming because of several converging lines of evidence, all pointing in the same direction. These include "an understanding of basic physics, comparing observations with models, and fingerprinting the detailed patterns of climate change caused by different human and natural influences," as stated in a 2014 report by the U.S. National Academy of Sciences and British Royal Society.

For instance, we have observed a 40% rise in CO_2 levels since the dawn of the industrial age in the mid-1700s, from 280 parts per million (ppm) in the air to approximately 400 ppm now. In 1970, CO_2 levels were around 325 ppm, which means that most of the rise in CO_2 from preindustrial levels has occurred in the last four decades, matching a time of rapid growth in industrial energy use and CO_2 emissions. In addition, the amount of total warming the planet has experienced since 1900 is inconsistent with the temperature change you would expect just from the observed natural changes, such as volcanic emissions and the decrease in solar radiation. In fact, as noted, absent human-caused greenhouse gases, we would expect the Earth to be in a period of global cooling right now. It is only by including all of human activity, factoring in all of the greenhouse gases emitted by fossil fuel consumption, that we see a consistency between observed temperature change and basic physics calculations.

There are numerous other "human fingerprints" on the pattern of climate changes we have observed in recent decades. When scientists have specifically measured the type of carbon (the particular ratio of carbon isotopes) building up in our atmosphere, most of it is exactly the type that scientists know comes from combustion of fossil fuels, as opposed to other sources, such as deforestation, which plays a much smaller role. The U.S. National Academy and United Kingdom's Royal Society note: "The observed patterns of surface warming, temperature changes through the atmosphere, increases in ocean heat content, increases in atmospheric moisture, sea level rise, and increased melting of land and sea ice also match the patterns scientists expect to see due to rising levels of CO_2 and other human-induced changes." In particular, climate science predicts that if the warming is caused by an increase in greenhouse gases, we expect the lower atmosphere (troposphere) to warm, the upper atmosphere (stratosphere) to cool, and the boundary between them (tropopause) to rise. All of this has been observed. If, for instance, recent warming were due to increases in the intensity of radiation from the sun, then in addition to the troposphere, the stratosphere should be warming, too, which is not happening.

What connects the greenhouse gases with the warming? The website Skeptical Science summarizes the research findings:[4]

- Satellites measure less heat escaping out to space, at the particular wavelengths that CO_2 absorbs heat, thus finding "direct experimental evidence for a significant increase in the Earth's greenhouse effect."
- If less heat is escaping to space, where is it going? It's going back to the Earth's surface. Surface measurements confirm this, observing more downward infrared radiation. A closer look at the downward radiation finds more heat returning at CO_2 wavelengths, leading to the conclusion that "this experimental data should effectively end the argument by skeptics that no experimental evidence exists for the connection between greenhouse gas increases in the atmosphere and global warming."

So we know it is humanity burning fossil fuels that is raising CO_2 levels. In addition, we know that this increased CO_2 is trapping heat precisely in the manner climatologists had long predicted. Moreover, we know the pattern of impacts from this warming are precisely what climate scientists predicted.

Finally, the confidence scientists have does not just come from the fact that every aspect of climate change in recent decades fits the precise pattern predicted from climate science for increases in human-caused greenhouse gases. At the same time, no alternative theory has ever been presented that can account for all of the observations. Moreover, such an alternative theory would not merely have to provide a mechanism to account for the warming and other observed changes, it would also have to come up with another as-yet unknown mechanism that was somehow negating the warming that science has long predicted from human-caused greenhouse gases.

Why has the climate changed in the past, before there were human-caused greenhouse gas emissions?

The major climate changes of the past all occurred because the climate was driven to change by some external change, which is typically called a climate forcing. These forcings include changes in the intensity of the sun's radiation, volcanic eruptions (which generally cause a short-term cooling), rapid releases of greenhouse gases, and changes in Earth's orbit.

In particular, the biggest climate changes in the past 800,000 years have been the ice-age cycle, "slow changes in Earth's orbit which alter the way the Sun's energy is distributed with latitude and by season on Earth," as the U.S. National Academy of Sciences and British Royal Society put it in 2014.

A key point about the global climate is that it does not appear to be inherently stable. As Wallace Broecker, a leading climatologist, wrote in the journal *Nature* in 1995, "The paleoclimate record shouts out to us that, far from being self-stabilizing, the Earth's climate system is an ornery beast which overreacts even to small nudges."[5]

Here, for instance, is the paleoclimate record of recent ice ages: an overlay of CO_2 levels in parts per million by volume (ppmv; Figure 1.2, top curve) over the past 800,000 years with the temperature in Antarctica during the same period (in °C, Figure 1.2, bottom curve) derived from various ice core samples. The trace gases that are found in deep ice layers reveal both temperatures and CO_2 levels.

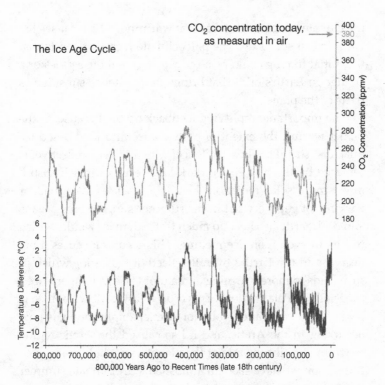

Figure 1.2 Historical CO_2 (top right axis) and reconstructed temperature (compared with the average temperature for the last 100 years) based on Antarctic ice cores for the last 800,000 years. Image via U.S. National Academy of Science.

The data reveal that when an initial warming is triggered by an external forcing (such as an orbital change), the planet can warm up at a fast rate. That in turn implies the climate system has strong amplifying feedbacks, which turn a small initial warming into a large heating fairly quickly.

What are the climate system's amplifying feedbacks that turn a moderate initial warming into a big ultimate warming?

The long-term historical record indicates that after some forcing event starts the warming process, amplifying feedbacks in the climate systems reinforce that warming, which causes the warming to speed up. The paleoclimate record suggests that the initial forcing could be a release of greenhouse gases or a change in Earth's orbit that brings more intense sunshine to parts of the planet.

One important amplifying feedback occurs because, as the planet warms, the extent of both sea ice and land-based ice (glaciers) shrinks. Thus white ice, which is very reflective, is replaced by the blue sea or dark land, each of which absorb much more solar radiation. Just as a black asphalt road or parking lot gets very hot in the summer sun, the blue oceans and dark earth also heat up much faster than ice would, which results in even more ice melting. This feedback causes a big decrease in the Earth's overall reflectivity (albedo), which in turn leads to more warming and a rapid rise in temperatures, especially in Polar regions. This crucial fast feedback, which is part of a process called polar (or Arctic) amplification, is now occurring in the Arctic, and it has caused the Arctic region to warm at twice the rate of the planet as a whole. It is a central reason we have seen an almost 80% drop in late summer Arctic ice volume since 1979 and a more than five-fold increase in the Greenland ice sheet melt rate in the past two decades.

Another key rapidly acting amplifying feedback is driven by water vapor. As the planet starts to heat up, evaporation increases, which puts more water vapor into the air. Water vapor is a potent heat-trapping greenhouse gas. So an increase in water vapor causes an increase in warming, which causes an increase in water vapor, and so on. A 2008 paper analyzing recent changes in surface temperature and the response of lower atmosphere water vapor to these changes concluded that the "water-vapor feedback implied by these observations is strongly positive" and "similar to that simulated by climate models."[6] The lead author, Professor Andrew Dessler, a climatologist at the Department of Atmospheric Sciences of Texas A&M University, has said that this finding is "unequivocal." That analysis concluded:

"The existence of a strong and positive water-vapor feedback means that projected business-as-usual greenhouse-gas emissions over the next century are virtually guaranteed to produce warming of several degrees Celsius. The only way that will not happen is if a strong, negative, and currently unknown feedback is discovered somewhere in our climate system."

To date, no such strong negative feedback that operates over a time scale of decades or less has been found. We find that the reverse is true: a number of additional strong positive feedbacks have been observed. The most important of these feedbacks involves the way warming, driven by greenhouse gases, can cause more greenhouse gases to be emitted from the Earth. For instance, to the extent that climate change leads to more forest fires, the CO_2 released by burning trees acts as an amplifying feedback, which then causes more climate change. More significantly, many studies have found that global warming can cause the soil or oceans or tundra (permafrost) to release additional CO_2 and CH_4, both strong greenhouse gases. These feedbacks can potentially drive up projected global temperatures a great deal this century and may ultimately determine just how destructive planetary warming is this century. For this reason, they will be discussed in greater detail in Chapter Three.

Is the current level of atmospheric CO_2 concentration unprecedented in human history?

Carbon dioxide levels in the air have now passed 400 parts per million (ppm). The U.S. National Academy of Sciences and United Kingdom Royal Society explained in 2014, "The present level of atmospheric CO_2 concentration is almost certainly unprecedented in the past million years, during which time modern humans evolved and societies developed." Anatomically modern humans, Homo sapiens sapiens, date back no more than 200,000 years. As Figure 1.2 shows, during that time—and going back a total of 800,000 years—CO_2 levels generally never exceeded 280–300 ppm.

The last time the Earth's atmosphere was at 400 ppm of CO_2 was a few million years ago, long before Homo sapiens roamed the Earth. Back then, the climate was 2°C (3.6°F) to 3°C (5.4°F) above preindustrial temperatures, and sea level was some 15–25 meters (50–80 feet) above modern levels. A 2009 analysis in *Science* found that when CO_2 levels were approximately 400 ppm 15 to 20 million years ago, the Earth was 5°F to 10°F warmer globally and seas were also 75 to 120 feet higher. So it is no surprise that current levels of CO_2 are leading to rapid warming and sea level rise.

It is not just the absolute CO_2 level that is unprecedented in the experience of modern humans. So is the *rate of change* of CO_2 levels. The rate matters for two reasons. First, the faster CO_2 levels change, the faster the planet warms up and the faster the climate changes, and thus the faster humans and other species must respond. We are currently headed toward climate change so rapid post-2050 that adaptation would become difficult if not impossible in many instances.

Second, there are some relatively slow processes (such as negative feedbacks) that can reduce CO_2 levels in the air over a time span of tens of thousands of years, keeping the Earth system in balance over very long periods of time. However, if CO_2 levels rise too fast, they overwhelm the ability of natural systems to absorb that CO_2. In fact, if CO_2 levels rise fast enough, the resulting warming and climate change can trigger amplifying feedbacks that cause natural systems to release more CO_2.

A 2008 *Nature Geosciences* study found we are currently releasing CO_2 into the atmosphere 14,000 times faster than nature has over the past 600,000 years, far too quickly for the slow, negative feedbacks to soak it up. The lead author concluded, "Right now we have put the system entirely out of equilibrium."

Are recent climatic changes unprecedented?

A stable climate enabled the development of modern civilization, global agriculture, and a world that could sustain a vast population, now exceeding 7 billion people. We already have unprecedented levels of CO_2 in the atmosphere, so it would not be surprising to learn that some of the CO_2-driven climate changes are unprecedented.

Until the last century, global temperatures over the past 11,000 years varied quite slowly, generally not more than a degree Fahrenheit (under a degree Celsius) over a period of several thousand years. In its final 2014 synthesis of more than 30,000 scientific studies, the Intergovernmental Panel on Climate Change concluded, "Warming of the climate system is unequivocal, and since the 1950s, many of the observed changes are unprecedented over decades to millennia." How unprecedented those changes were became clear in an earlier 2012 study, the most comprehensive scientific reconstruction of global temperatures over the past 11,000 years ever made. The study's funder, the National Science Foundation, explained in a news release: "During the last 5,000 years, the Earth on average cooled about 1.3 degrees Fahrenheit–until the last 100 years, when it warmed about 1.3 degrees F."[7] In short, primarily because of human-caused greenhouse gases, the global temperature is changing 50 times faster than it did during the time when modern civilization and agriculture developed, a time when humans figured out where the climate conditions—and rivers and sea levels—were most suited for living and farming.

In 2013, scientists from the International Programme on the State of the Ocean reported that the rate we are acidifying the oceans is also "unprecedented." Approximately one quarter of the CO_2 humans emit into the air gets absorbed in the oceans. The CO_2 that dissolves in seawater forms carbonic acid, which in turn acidifies the ocean. As a result, the oceans are more acidic today than they have been over the last 300 million years. A 2010 study concluded that the oceans are acidifying 10 times faster today than 55 million years ago when a mass extinction of marine species occurred (see Chapter Three).

Has recent human-caused climate change been occurring faster or not as fast as scientists predicted?

A great many of the impacts from human-caused climate change have been occurring within the range that scientists had projected. Some of the most important impacts from climate change have been occurring considerably faster than scientists expected.

Consider the Arctic ice cap. After 2000, the Arctic began to lose sea ice several decades ahead of every single climate model the Intergovernmental Panel on Climate Change (IPCC) was using at the time. Those models had projected that the Arctic Ocean would not go ice free in the summer until 2080 or later. However, from 1979 to 2012, late-summer Arctic sea ice volume dropped by 80%.

Likewise, early this century, scientists did not expect that the great ice sheets of Greenland and Antarctica would melt enough to contribute much to total sea level rise by 2100. However, observations revealed unexpectedly fast melt. The latest observations suggest they will be a primary, if not the primary, driver of sea level.

A 2012 study, "Comparing climate projections to observations up to 2011," confirmed that climate change is happening as fast—and in some cases faster—than climate models had projected. The *Environmental Research Letters* study found, "The rate of sea-level rise in the past decades is greater than projected by the latest assessments of the IPCC, while global temperature increases in good agreement with its best estimates." In particular, the oceans are rising 60% faster than the IPCC's latest best estimates. The news release notes, "The increased rate of sea-level rise is unlikely to be caused by a temporary episode of ice discharge from the ice sheets in Greenland or

Antarctica or other internal variabilities in the climate system, according to the study, because it correlates very well with the increase in global temperature." Lead author Stefan Rahmstorf from the Potsdam Institute for Climate Impact Research said, "The new findings highlight that the IPCC is far from being alarmist and in fact in some cases rather underestimates possible risks."

Is there a difference between global warming and climate change?

Global warming generally refers to the observed warming of the planet due to human-caused greenhouse gas emissions. Climate change generally refers to all of the various long-term changes in our climate, including sea level rise, extreme weather, and ocean acidification.

In 1896, a Swedish scientist, Svante Arrhenius, concluded that if we double atmospheric CO_2 levels to 560 ppm (from preindustrial levels of 280), then surface temperature levels would rise several degrees. The first published use of the term "global warming" appears to have been in 1975 by the climatologist Wallace Broecker in an article in the journal *Science* titled, "Climatic Change: Are We on the Brink of a Pronounced Global Warming?" In June 1988, *global warming* became the more popular term after NASA scientist James Hansen told Congress in a widely publicized hearing that "Global warming has reached a level such that we can ascribe with a high degree of confidence a cause and effect relationship between the greenhouse effect and the observed warming."

The term "climate change" dates at least as far back as 1939. A closely related term, "climatic change," was also common, as in the 1955 scientific article, "The Carbon Dioxide Theory of Climatic Change" by Gilbert Plass. By 1970, the journal *Proceedings of the National Academy of Sciences* published a paper titled "Carbon Dioxide and its Role in Climate Change." When the world's major governments set up an advisory body in 1988 of top scientists and other climate experts to review the scientific literature every few years, they named it the "Intergovernmental Panel on *Climate Change.*"

Climate change or global climate change is generally considered a "more scientifically accurate term," than global warming, as NASA explained in 2008, in part because "Changes to precipitation patterns and sea level are likely to have much greater human impact than the higher temperatures alone."[8] When you consider all of the impacts scientists have observed in recent decades—including the acidifying ocean, worsening wildfires, and more intense deluges—climate scientists are likely to continue favoring the term climate change. In general or popular usage, global warming and climate change have become interchangeable over the past several decades, and that trend is likely to continue this century, especially as the warming itself becomes more and more prominent.

What are the sources of the most important human-caused pollutants that drive global warming?

The primary greenhouse gas generated by human activity is CO_2. In the United States, for instance, in 2012, CO_2 accounted for 82% of U.S. greenhouse gas emissions, primarily from burning fossil fuel hydrocarbons (i.e., coal, oil, and natural gas). Of all human-caused CO_2, more than 90% comes from burning fossil fuels (coal, oil, and natural gas) and cement making. The rest of human-caused CO_2 comes from land-use change, especially deforestation. In 2012, CH_4 comprised up to 9% of U.S. greenhouse gases. Natural gas is mostly CH_4. Major sources of CH_4 include leaks during the extraction and transportation of fossil fuels, livestock (like cows), decaying organic waste in landfills, and some agricultural practices. Nitrous oxide (N_2O) made up 6% of U.S. greenhouse gas emissions. Major sources of N_2O include agriculture and combustion of fossil fuels and solid waste.[9]

Another key pollutant that drives global warming is black carbon. Black carbon makes up a major proportion of soot and fine particulate matter. It is very highly light absorbing. In the United States, slightly more than half of black carbon emissions come from transportation (mobile sources), and of that, more than 90% comes from diesel engines. The other large U.S. source is the burning of biomass, including wildfires. Biomass burning is the largest single source of global black carbon emissions. Because black carbon is so highly light absorbing, it directly changes the amount of solar radiation absorbed by the atmosphere and land. In particular, black carbon that has been deposited on snow and ice reduces their reflectivity (albedo), which means they absorb more sunlight and therefore boost the temperature and melt rate in places like Greenland, Antarctica, and the Arctic.

How does deforestation contribute to warming?

Trees and plants take CO_2 out of the air and emit oxygen. This is part of the photosynthesis process whereby trees and plants convert sunlight into energy. Vegetation is thus a "carbon sink," causing a net reduction in atmospheric CO_2 (as opposed to a "carbon source" such as fossil fuel combustion).

Deforestation reduces the carbon sink, and the decay of the resulting dead plant matter actually becomes a new source of carbon. In many cases, deforestation is accompanied by burning the dead trees and plant matter, which releases most of the carbon that had been stored in them. In its 2007 assessment of the scientific literature on climate, the IPCC concluded that deforestation was responsible for 17% of all greenhouse gas emissions, with most of those emissions coming from the destruction of tropical forests in places such as Brazil and Indonesia.

In the past decade, however, Brazil sharply reduced its rate of deforestation. Brazil reduced its annual rate of Amazon deforestation by 80% between 2004 and 2013 (although the rate increased in 2014 and 2015). At the same time, global emissions of CO_2 from burning fossil fuels soared, with China as the lead contributor. The net result is that today, deforestation is responsible for closer to 8% of all greenhouse gas emissions, according to the Global Carbon Project.

What is global warming potential and why is it different for various greenhouse gases?

Different greenhouse gases trap different amounts of heat in the atmosphere. The global warming potential (GWP) compares how much heat a greenhouse gas traps compared to a similar mass of CO_2. Because different greenhouse gases have different lifetimes (last different lengths of time in the air), the GWP also generally varies by time.

For instance, methane (CH_4) is a far more potent greenhouse gas than CO_2, especially over shorter periods of time. In 2013, the Intergovernmental Panel on Climate report reported that CH_4 has 34 times stronger a heat-trapping gas than CO_2 over a 100-year time scale, so its 100-year GWP is 34. The Panel reported that, over a 20-year time frame, CH_4 has a GWP of 86 compared with CO_2. A large part of the difference is that the atmospheric lifetime of CH_4 is approximately 12 years, whereas the lifetime of CO_2 in the air is far longer. Some of the CO_2 that humans are putting into the air stays there for thousands of years. Scientists and governments seeking to reduce future warming have focused on CO_2 not merely because of the vast quantity we are putting in the air, but also because of its astonishingly long lifetime once it gets there.

Historically, the 100-year GWP has been by far the most widely used in studies of future climate change. The reason for this emphasis is because scientists and governments have been focused on the long-term warming trend and related impacts we will see by the end of the 21st century. However, given how close we seem to be getting to certain irreversible tipping points, some scientists have argued that we should use a shorter time span, such as 20 years. In its major 2013 literature review, the IPCC concluded the following: "There is no scientific argument for selecting 100 years compared with other choices. The choice of time horizon is a value judgement since it depends on the relative weight assigned to effects at different times."

Why does the rate of warming appear to vary from decade to decade?

The rate of global warming has not been constant over the past century—if we measure warming by the change in surface temperatures. There have been periods lasting a decade or more where the rate of surface warming has been very fast, and there have been periods when it has been relatively slow.

This variation is primarily due to a variety of natural and human-caused "forcings" that serve to temporarily speed up and slow down the overall warming trend. These include the solar cycle, particulates (sulfate aerosols) from both volcanoes and human-caused pollution, and the El Niño Southern Oscillation, also known as the El Niño–La Niña cycle.

An El Niño is a relatively short-term climate event in which the Equatorial Pacific sees warmer than average ocean temperatures, whereas in a La Niña, we see colder-than-normal temperatures in the same region. Both events are associated with extreme weather around the globe. El Niños are generally the hottest years on record, because the regional warming adds to the underlying global warming trend. La Niña years tend to be below the global warming trend line.

Has global warming slowed down or paused in recent years?

Human-caused greenhouse gases keep trapping more and more heat. The rate of heat build-up for the entire planet is 250 trillion watts, a large and abstract number. A number of scientists have made the analogy to the energy released by the Hiroshima atomic bomb. By that metric, the current rate of increase in global warming is roughly the same as detonating 400,000 Hiroshima bombs per day, 365 days per year. This is a vast amount of warming, which has sped up, not slowed down.

We know that more than 90% of global warming goes into the oceans, whereas only 1% goes into the atmosphere. So we would expect measurements of ocean warming to be the most reliable indicator of global warming, because a relatively small amount of atmospheric warming can be temporarily swamped by other forcings. As discussed earlier, recent studies have found the upper ocean (above 700 meters or 2300 feet) to be warming much faster in the past four decades than previously thought, and that "recent warming rates of the waters below 700m appear to be unprecedented" and speeding up, which has been "contributing significantly to an acceleration of the warming trend." We have also seen accelerated loss of ice on the Arctic sea ice, Greenland ice sheet, and Antarctic ice sheet, which also signals that overall global warming is speeding up.

The question remains that, although total planetary warming obviously has continued unabated, has the rate of rise of surface temperatures slowed down (or even stopped) in recent years, and if so why? The answer is no. The rise of surface air temperatures has not stopped. The apparent recent (temporary) slow down in the rate of surface warming was largely due to the natural and human-caused forcings that temporarily speed up and slow down the overall warming trend. That seeming slow down has ended.

Global temperature data are tracked by various groups around the world, including NASA, NOAA, Japan's meteorological agency, and the United Kingdom's Met Office. It has primarily been in the United Kingdom's Met Office data that one finds evidence of an extended recent stop or hiatus in warming. Why is that? "There are no permanent weather stations in the Arctic Ocean, the place on Earth that has been warming fastest," *New Scientist* has explained. "The UK's Hadley Centre record simply excludes this area, whereas the NASA version assumes its surface temperature is the same as that of the nearest land-based stations." That is one reason we know with high certainty that the planet has actually warmed up more in the past decade than reported by some of the global temperature records, especially the Met Office, which uses "HadCRUT" data developed by the Hadley Center with the Climate Research Unit ([CRU] Norwich, UK).

In December 2013, researchers showed that these "missing" data had caused a large part of the supposed slowdown in the Met office data.[10] German Climatologist Stefan Rahmstorf summarized the findings this way:

A new study by British and Canadian researchers shows that the global temperature rise of the past 15 years has been greatly underestimated. The reason is the data gaps in the weather station network, especially in the Arctic. If you fill these data gaps using satellite measurements, the warming trend is more than doubled in the widely used HadCRUT4 data, and the much-discussed "warming pause" has virtually disappeared.

When you include all of the data scientists have (through 2012), surface air temperatures have continued to rise globally in the last decade (see Figure 1.3), but at what appears to be a slightly slower rate than in previous decades. Why is that? A 2011 study removed the "noise" of natural climate variability from the temperature record to reveal the true global warming signal.[11] That noise is "the estimated impact of known factors on short-term temperature variations (El Niño/southern oscillation, volcanic aerosols and solar variability)." When they did that, researchers found "the warming rate is steady over the whole time interval" from 1979 through 2010. A 2012 study by the Potsdam Institute for Climate Impact Research led by Rahmstorf found "The rate of sea-level rise in the past decade is greater than projected by the latest assessments of the IPCC, while global temperature increases in good agreement with its best estimates." On the subject of global warming, Rahmstorf explains, "Global temperature continues to rise at the rate that was projected in the last two IPCC Reports. This shows again that global warming has not slowed down or is lagging behind the projections." The study averages five global temperature series and compared them to the IPCC:

> To allow for a more accurate comparison with projec-
> tions, the scientists accounted for short-term tempera-
> ture variations due to El Niño events, solar variability
> and volcanic eruptions. The results confirm that global
> warming, which was predicted by scientists in the 1960s
> and 1970s as a consequence of increasing greenhouse
> concentrations, continues unabated at a rate of 0.16°C per
> decade and follows IPCC projections closely.

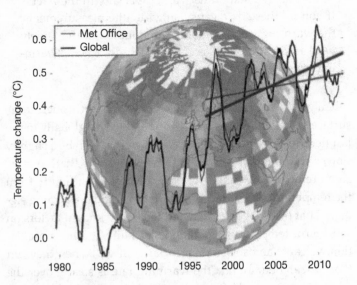

Figure 1.3 The corrected data (bold lines) are shown compared to the uncorrected ones (thin lines).

Source: Kevin Cowtan and Robert Way

A 2013 study published in *Nature* confirmed "the slowing rise in global temperatures during recent years has been a result of prevalent La Niña periods in the tropical Pacific." Thus, there are, as Rahmstorf notes, "at least three independent lines of evidence that confirm we are not dealing with a slowdown in the global warming trend, but rather with progressive global warming with superimposed natural variability."

In January 2015, Dr. Gavin Schmidt, director of NASA's Goddard Institute of Space Studies, tweeted, "Is there evidence that there is a significant change of trend from 1998? (Spoiler: No)." He attached this chart (Figure 1.4), which uses NASA's latest data.

A June 2015 *study in Science* from a team of NOAA scientists confirms "Data show no recent slowdown in global warming." As NOAA explains, observations reveal that "the rate of global warming during the last 15 years has been as fast as or faster than that seen during the latter half of the 20th Century."

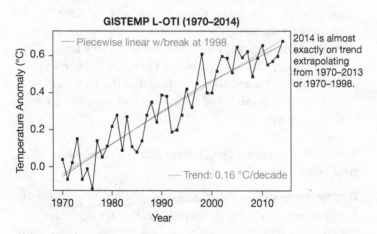

Figure 1.4 The latest NASA temperature data make clear that not only has there been no "pause" in surface temperature warming in the past decade and a half, there has not even been a significant change in trend.

Can we reach a point where emitting more CO_2 into the air will not cause more climate change?

Adding additional CO_2 to the air will always cause more global warming. As CO_2 levels rise, additional CO_2 becomes less effective at trapping heat; however, warming still increases. This "understanding of the physics by which CO_2 affects Earth's energy balance is confirmed by laboratory measurements, as well as by detailed satellite and surface observations of the emission and absorption of infrared energy by the atmosphere," as noted by a 2014 report by the U.S. National Academy of Sciences and British Royal Society. That report explains:

> Greenhouse gases absorb some of the infrared energy that Earth emits in so-called bands of stronger absorption that occur at certain wavelengths. Different gases absorb energy at different wavelengths. CO_2 has its strongest heat-trapping band centred at a wavelength of 15 micrometres (millionths of a metre), with wings that spread out a few micrometres on either side. There are also many weaker absorption bands. As CO_2 concentrations increase, the absorption at the centre of the strong band is already so intense that it plays little role in causing additional warming. However, more energy is absorbed in the weaker bands and in the wings of the strong band, causing the surface and lower atmosphere to warm further.

Have we already crossed tipping points (points of no return) in the climate system?

The latest science suggests that we are getting close to levels of greenhouse gases in the atmosphere that will trigger irreversible changes, and, in at least once case, we may have already crossed a tipping point.

In 2009, a team of researchers led by NOAA scientists published a major study that concluded: "the climate change that is taking place because of increases in CO_2 concentration is largely irreversible for 1,000 years after emissions stop."[12] This NOAA-led study found that some of the most severe long-term impacts, such as drops in precipitation and unstoppable sea level rise, would become irreversible this century if CO_2 levels continue to rise as they have (because of human activity):

> Among illustrative irreversible impacts that should be expected if atmospheric CO_2 concentrations increase from current levels near 385 parts per million by volume (ppm) to a peak of 450-600 ppm over the coming century are irreversible dry-season rainfall reductions in several regions comparable to those of the "dust bowl" era and inexorable sea level rise.

We are now near 400 ppm and rising more than 2 ppm a year. We are headed to CO_2 levels far above 600 ppm this century on our current emissions trajectory. A key point is that irreversible does not mean unstoppable, especially if we can keep total warming below 2°C (3.6°F), which is roughly an atmospheric concentration of CO_2 of 450 ppm.

That said, two studies from 2014 find that some Western Antarctic ice sheet glaciers "have passed the point of no return," according to Eric Rignot, the lead author of one of the 2014 studies. Rignot, a glaciologist for NASA and University of California at Irvine said, "The collapse of this sector of West Antarctica appears to be unstoppable." Such a collapse by itself would raise sea levels 4 feet in the coming centuries. Crucially, these glaciers act "as a linchpin on the rest of the [West Antarctic] ice sheet, which contains enough ice to cause" a total of 12 to 15 feet of global sea level rise.

Commenting on these new studies, sea-level-rise expert Stefan Rahmstorf, Co-Chair of Earth System Analysis, Potsdam Institute for Climate Impact Research, wrote:

What climate scientists have feared for decades is now beginning to come true: We are pushing the climate system across dangerous tipping points. Beyond such points things like ice sheet collapse become self-sustaining and unstoppable, committing our children and children's children to massive problems. The new studies strongly suggest the first of these tipping points has already been crossed. More tipping points lie ahead of us. I think we should try hard to avoid crossing them.

2

EXTREME WEATHER AND CLIMATE CHANGE

*Extreme weather is the earliest manifestation of climate change
that most people will be exposed to. This chapter will focus on
what science tells us about the remarkable spate of once-in-a-
century (and rarer) extreme weather events we have observed
in the last few years. It will emphasize what we can—and cannot—say about their connection to climate change.*

What is the difference between weather and climate?

"Climate is what you expect; weather is what you get." That
saying (variously attributed to Mark Twain, the science fiction
writer Robert Heinlein, and others) captures the distinction.
The weather is the set of atmospheric conditions you experience at a specific time and place. Is it hot or cold? Is it raining or dry? Is it sunny or cloudy? The climate is the statistical
average of these weather conditions over a long period of time,
typically decades. Is it a tropic climate or a polar climate? Is it
a rainforest or a desert?

Why is long-term weather prediction very hard? Because on
any given day—1 year from now or 10 years from now—the
possible temperature range spans tens of degrees Fahrenheit
or even Celsius. Likewise, there could be a deluge or no rain at
all on any given day.

The climate is considerably easier to predict precisely
because it is a long-term average. Greenland is going to be

much colder than Kenya during the course of a year and during almost every individual month. The Amazon is going to be much wetter than the Sahara desert virtually year-round.

We call a weather event "extreme" when it is far outside the normal climate for that location and time of year, especially if that event extends over many days or even months and if it covers a vast area. If almost all of Greenland is unusually warm for a month or if virtually the entire Amazon is unusually dry for a month, those are extreme weather events.

How extreme or rare a weather event is will often be described in terms of how often it reoccurs—once every 10 years, once every 100 years, once every 1000 years. Although the climate is supposed to be a statistical average that changes little if at all over short periods of time such as decades, we are now rapidly changing the climate, creating what is often called a "new normal." Storms that were previously 100-year storms are becoming 10-year storms.

Because climate change is expected to make dry or semi-arid regions hotter and drier, we would expect longer and more intense droughts in such regions, such as the Mediterranean and U.S. Southwest. Eventually, the climate is projected to change so much that these regions' normal climate becomes a drought.

Which extreme weather events are being made worse by climate change and which are not?

Warming directly makes heat waves longer, stronger, and more frequent. For instance, a major 2012 study found that extreme heat waves in Texas, such as the one that occurred in 2011, are much more likely—20 times more likely in years like 2011—to occur than they were 40–50 years ago.[13]

Although human-caused global warming makes extremely warm days more likely, it makes extremely cold days less likely. So while we will continue to have record-setting cold temperatures in places, the ratio of record-setting hot days to record-setting cold days will grow over time, which has been measured. The U.S. National Center for Atmospheric Research (NCAR) reported in late 2009 that "Spurred by a warming climate, daily record high temperatures occurred twice as often as record lows over the last decade across the continental United States, new research shows." Likewise, the UK Met Office reported in 2014 that, globally, the ratio of days that are extremely warm versus the days that are extremely cold has risen sharply since 1950. They point out "Globally, 2013 was also in the top 10 years for the number of warm days and in the bottom 10 years for the number of cool nights since records began in 1950."

Global warming directly makes droughts more intense by drying out and heating up land that is suffering from reduced precipitation. The warming also worsens droughts by causing earlier snowmelt, thus reducing a crucial reservoir used in the West during the dry summer season. Finally, climate change shifts precipitation patterns, causing semi-arid regions to become parched. For instance, the 2012 Texas study found "indications of an increase in frequency of low seasonal precipitation totals."

The heat and the drying and the early snow melt also drive worsening wildfires, particularly in the West. The wildfire season is already more than 2 months longer than it was just a few decades ago, and wildfires are much larger and more destructive.

Warming also puts more water vapor in the atmosphere, so that wet areas of the world become wetter and deluges become more intense and more frequent. This effect has already been documented and linked to human activity in the northern hemisphere. As New York Governor Andrew Cuomo said after Superstorm Sandy slammed his state just 2 years after it was deluged by hurricane Irene, "We have a one-hundred year flood every two years now." Note that this means that when it is cold enough to snow, snowstorms will be fueled by more water vapor and thus be more intense themselves. We thus expect fewer snowstorms in regions close to the rain-snow line, such as the central United States, although the snowstorms that do occur in those areas are likely to be more intense. It also means we expect more intense snowstorms in generally cold regions. This may appear to be counterintuitive, but the warming to date is not close to that needed to end below-freezing temperatures over large parts of the globe, although it is large enough to put measurably more water vapor into the air.

In addition, warming raises sea levels by heating up and expanding water and by melting landlocked ice in places such as Greenland and Antarctica. Those rising sea levels in turn make devastating storm surges more likely. For instance, warming-driven sea level rise nearly doubled the probability of a Sandy-level flood today compared with 1950. Studies also find that global warming makes the strongest hurricanes more intense, because hurricanes draw their energy from ocean warmth, so that once a hurricane forms, global warming provides it more fuel. The question of how global warming affects tornadoes formation is very complicated and will be addressed later in this chapter.

What is the role of natural climatic variation, such as the El Niño–La Niña cycle, in extreme weather?

"We expect some of the most significant impacts of climate change to occur when natural variability is exacerbated by long-term global warming, so that even small changes in global temperatures can produce damaging local and regional effects." That was part of a 2009 statement by the Met Office (the United Kingdom's National Weather Service) and the Royal Society (the UK's National Academy of Science). The largest short-term contributor to the "natural dynamical variability" of the climate system is the El Niño–La Niña cycle, which was discussed in Chapter One. Many of the most extreme events we see today are associated with the combination of human-caused warming and either El Niño or La Niña.

At the time, 2010 was the hottest year on record, driven by global warming and a moderate El Niño. Meteorologist Dr. Jeff Masters said 2010 was "the planet's most extraordinary year for extreme weather since reliable global upper-air data began in the late 1940s"—and quite possibly since 1816 (the infamous "Year Without a Summer," which itself was driven by the largest volcanic eruption in over a millennium, the 1815 eruption of Indonesia's Mount Tambora).[14]

Consider what kind of extreme events the world saw with this El Niño added to global warming. First, 2010 saw 20 countries set all-time record highs—the most records ever set in 1 year—including "Asia's hottest reliably measured temperature of all-time, the remarkable 128.3°F (53.5°C) in Pakistan in May 2010." The Arctic saw its atmospheric circulation take on its "most extreme configuration in 145 years of record keeping." One result is that Canada had its warmest winter on record, shattering many all-time temperature records. That required, for the first time, Canadian officials to have to helicopter in snow for the 2010 Vancouver Olympics. In January 2010, the U.S. Southwest from California to Arizona was slammed by "The most powerful low pressure system in 140 years of record keeping." That system brought hurricane force winds exceeding 90 miles per hour, enormous dust storms, tornadoes, and blizzards. In May 2010, Tennessee was deluged by an unprecedented superstorm. For much of the western half of the state, this was a once in 500-year flood. For the Nashville area, it was a rarer than once in a 1000-year flood, driven by more than 13 inches of rain over a 2-day period—a deluge that exceeded the previous record of 11 inches for total rainfall in the entire month of May.

That summer, Russia was hit by the most lethal heat wave in human history, killing at least 55,000 people. Masters notes that "Moscow's old extreme heat record, 37°C (99°F) in 1920, was equaled or exceeded *five times in a two-week period* from July 26 to August 6 2010, including an incredible 38.2°C (101°F) on July 29." In August 2010, the head of the Russian

Meteorological Center said, "We have an 'archive' of abnormal weather situations stretching over a thousand years. It is possible to say there was nothing similar to this on the territory of Russia during the last one thousand years in regard to the heat." Russia lost 40% of its wheat crop and banned grain exports for 18 months, which contributed to soaring food prices globally, which in turn set the stage for unrest in the Middle East and around the world. Moreover, during that summer, Pakistan was hit by the costliest natural disaster in its history, a deluge that put more than one fifth of the country underwater, affecting some 20 million people. In addition, in 2010, both Columbia and Australia saw their worst floods in history, driven by record rainfall. In October 2010, Minnesota saw the strongest U.S. storm ever recorded that was not a coastal storm such as a hurricane. The superstorm generated 67 tornadoes over a period of 4 days.

In 2010, the Amazon experienced its second 100-year drought in 5 years, killing a large number of rain forest trees and causing tremendous emissions of carbon dioxide. A study in the journal *Science*, "The 2010 Amazon Drought," concluded that, if this pattern continues, the Amazon rain forest as we have come to know it will ultimately be severely degraded or destroyed. That would turn one of the Earth's carbon "sinks," a place that stores vast amounts of carbon dioxide, into a "source" of carbon dioxide. Finally, in December 2010, Greenland saw "the strongest ridge of high pressure ever recorded at middle levels of the atmosphere, anywhere on the globe (since accurate records began in 1948)."

Masters notes that any of these extreme weather events "could have occurred naturally sometime during the past 1,000 years." However, he adds, it is "highly improbable that the remarkable extreme weather events of 2010"—and additional extreme events in the whole first half of 2011—"could have all happened in such a short period of time without some powerful climate-altering force at work. The best science we have right now maintains that human-caused emissions of heat-trapping gases like CO_2 are the most likely cause of such a climate-altering force." Therefore, it is the unprecedented run of extreme events—each individually unprecedented—that provides a strong set of fingerprints that human-caused climate change is already having a noticeable impact on our weather.

Did climate change cause Hurricane Sandy (and why is that the wrong question to ask)?

"The answer to the oft-asked question of whether an event is caused by climate change is that it is the wrong question," as climatologist Kevin Trenberth wrote in the journal *Climatic Change*. His 2012 paper, "How To Relate Climate Extremes to Climate Change," goes on to explain that "All weather events are affected by climate change because the environment in which they occur is warmer and moister than it used to be."

Climate change is making a variety of the most dangerous extreme weather events more extreme. It is also making them more likely and thus more frequent. As Meteorologist Dr. Jeff Masters said on the PBS *NewsHour* in December 2011: "We look at heat waves, droughts, and flooding events. They all tend to get increased when you have this extra energy in the atmosphere," from human-caused global warming. Like an athlete on steroids or performance-enhancing drugs, our climate system is breaking records at an unnatural pace. Masters explained that if you have a slugger who normally gets a big home run total, then "you add a little bit of extra oomph to his swing by putting him on steroids, now we can have an unprecedented season, a 70 home run season. And that's the way I look at this year." In addition, like a baseball player on steroids, it is the wrong question to ask whether a given home run is "caused" by steroids. Likewise, it is the wrong question to ask whether a given extreme weather event was "caused" by global warming.

Consider one of the most off-the-charts weather events in U.S history, Superstorm Sandy. On October 29, 2012, Hurricane Sandy devastated the Northeastern United States, killing more than 100 people, destroying entire communities, and inflicting more than $70 billion in damages. Sandy was the second costliest storm in U.S. history after 2005's Hurricane Katrina. Meteorologists explained it was the "largest hurricane in Atlantic history measured by diameter of gale force winds (1,040mi)." The National Weather Service called this "A Storm Like No Other" and pointed out: "I cannot recall ever seeing model forecasts of such an expansive areal wind field with values so high for so long a time. We are breaking new ground here." Stu Ostro, Weather Channel Senior Meteorologist, said: "History is being written as an extreme weather event continues to unfold, one which will occupy a place in the annals of weather history as one of the most extraordinary to have affected the United States." Ostro elaborated on what made Sandy so unique:

> A meteorologically mind-boggling combination of ingredients coming together: one of the largest expanses of tropical storm (gale) force winds on record with a tropical or subtropical cyclone in the Atlantic or for that matter anywhere else in the world . . . a "warm-core" tropical cyclone embedded within a larger, nor'easter-like circulation; and eventually tropical moisture and arctic air combining to produce heavy snow in interior high elevations. **This is an extraordinary situation, and I am not prone to hyperbole.**

Like many highly destructive extreme weather events, Sandy was caused by "a meteorologically mind-boggling combination of ingredients." That is a key reason it is the wrong question to ask whether Sandy was caused by climate change. There was a confluence of factors that caused its unique level of destruction. Significantly, though, climatologists have explained that global warming made Sandy more destructive in several ways:

1. **Warming-driven sea-level rise makes storm surges more destructive**. Human-caused climate change added nearly one foot to the total Sandy storm surge, which in turn exposed another 25 square miles and 40,000 people to flooding. A 2012 study by the U.S. Geological Survey found the rate of sea-level rise has been up to four times faster than the global average along parts of the Atlantic Coast including New York, Norfolk, and Boston.[15]

2. **Global warming makes deluges more intense**. Higher sea surface temperatures mean additional water vapor in the atmosphere, which produces 5 to 10% more rainfall, which in turn raises the risk of flooding. The bigger the storm, the more additional moisture swept into it thanks to global warming.

3. **Also, since warm water helps fuel hurricanes, warming makes the biggest storms more intense and bigger.** Relatedly, warming also extends the range of warm sea surface temperatures, which can help sustain the strength of a hurricane as it steers on a northerly track into cooler water. September 2012 had the second highest global ocean temperatures on record, and the Eastern seaboard of the United States was 5°F (2.8°C) warmer than average (with global warming responsible for at least one fifth of that extra warmth).

4. **The unusual path of the storm.** The storm track made, in Ostro's words, "a sharp left turn in direction of movement toward New Jersey in a way that is *unprecedented in the historical database*, as it gets blocked from moving out to sea by a pattern that includes an exceptionally strong ridge of high pressure aloft near Greenland." The sharp turn that directed Sandy into the heavily populated U.S. east coast rather than out to sea was caused by a very strong high-pressure system—the kind of "blocking pattern" that many recent studies have linked to warming.

I have put these in order from most scientific certainty to least. The first two—the impact of sea-level rise and increased water vapor—are unequivocal. The third is extremely likely. The fourth is more speculative. So human-caused global warming did not "cause" hurricane Sandy, but it certainly made the storm more damaging, and it may well be a key reason it ravaged coastal New York and New Jersey.

How does climate change affect heat waves?

Global warming raises the average temperature of the Earth over time. This makes heat waves, which are extremes on top of the average, more intense and more frequent. For the same reason, heat waves will last longer and cover a larger region.

However, a small shift in average temperatures can have a disproportionately large (or nonlinear) impact on how many people are exposed to the most extreme heat waves. As climatologists Stefan Rahmstorf and Dim Coumou of the Potsdam Institute for Climate Impact Research put it, "the same amount of global warming boosts the probability of *really* extreme events, like the [2012] US heat wave, far more than it boosts more moderate events."[16]

A detailed climatological analysis of historical global temperature data by NASA scientists in 2012 shows how this has already started happening. The NASA Goddard Institute for Space Studies researchers use the analogy of loaded "climate dice" to describe what humans are doing to weather extremes, particularly the chances for an unusually warm or cool summer. During the base historical period from 1951 to 1980, you could imagine those six-sided dice "with two sides colored red for 'hot', two sides blue for 'cold', and two sides white for near average temperatures." Each time you threw those dice, the chances of a hot summer or cold summer or near normal summer were roughly the same.

Those dice have become increasingly "loaded" to favor hot summers during the past three decades, the time of the most rapid human-caused global warming. The distribution of summer extremes has shifted towards higher temperatures, and the range of those extremes, especially on the hot side, has increased. Today, unusually "cool summers now cover only half of one side of a six-sided die, white covers one side, red covers four sides, and an extremely hot (red-brown) anomaly covers half of one side." Therefore, hot summers occur twice as often as they did, and cool summers occur far less often than they did. In addition, we now have a category of super-hot summers—those devastating heat waves that disproportionately harm to humans and animals and crops—that occur about as frequently as cool summers do.

More technically speaking, the researchers "illustrate variability of seasonal temperature in units of standard deviation (σ), including comparison with the normal distribution ('bell curve') that the lay public may appreciate." In particular, they look at a "subset of the hot category, extremely hot outliers, defined as anomalies exceeding $+3\sigma$," which "normally occur with a frequency of about 0.13%"—heat waves that occur in a location less than once a century. Historically, in a typical summer in the 1951–1980 period, "only 0.1-0.2% of the globe is covered by such hot extremes." However, their analysis found that this type of previously rare monster heat wave "now typically covers about 10% of the land area" during the summer months, as Figure 2.1 shows.

In short, the most extreme and most dangerous heat waves have seen a 50-fold increase. This increase is so dramatic, the NASA researchers conclude, "the extreme summer climate anomalies in Texas in 2011, in Moscow in 2010, and in France in 2003 almost certainly would not have occurred in the absence of global warming with its resulting shift of the anomaly distribution."

A few years ago, few climate scientists were willing to make such a strong statement of attribution, connecting a super extreme event, such as a once-in-1000-years heat wave, directly to human-caused global warming, and, even in 2014,

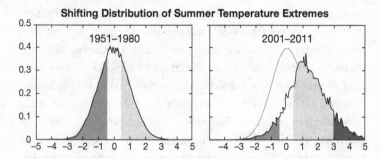

Figure 2.1 Frequency of occurrence (vertical axis) of local June–July–August temperature anomalies (relative to 1951–1980 mean) for Northern Hemisphere land in units of local standard deviation (horizontal axis). Temperature anomalies in the period 1951–1980 match closely the normal distribution ("bell curve"), which is used to define cold, typical, and hot seasons, each with probability 33.3% (via NASA).

not every climate scientist is ready to make such a strong statement. However, the jump in off-the-charts weather extremes in the past few years has led to an explosion of studies explaining how global warming is making those extremes considerably more frequent and destructive.

One last point on extreme heat waves. Like Superstorm Sandy (and other such large-scale events), the deadliest heat waves tend to be driven by "a meteorologically mind-boggling combination of ingredients." For instance, the very worst heat waves are typically driven in part by extreme drought, as discussed in the next answer.

How does climate change affect droughts?

A basic prediction of climate science is that many parts of the world will experience longer and deeper droughts, thanks to the combined effects of drying, warming, and the melting of snow and ice. In recent years, scientists have observed an increase in drought intensity and/or frequency due to global warming in many parts of the world.[17]

First, human-caused climate change has long been predicted to shift rainfall patterns and expand the dry regions of the world to encompass semi-arid regions, such as are found in the U.S. Southwest and the Mediterranean. The driest regions, including the major deserts, are found in the subtropics, two belts just outside the tropics (north and south of the equator). Climate science predicted these subtropical belts would expand, and they are. As a result, semi-arid regions become more drought prone. There is also emerging evidence that climate change makes some weather patterns, including droughts, more likely to get stuck or blocked by large high-pressure systems called blocking patterns. This may be part of the reason the California drought of 2012–2015 has been so long lasting.

Second, global warming causes greater evaporation and, once the ground is dried out, the Sun's energy goes into baking the soil, leading to a further increase in air temperature. That is why, for instance, so many temperature records were set for the United States in the 1930s Dust Bowl, and why, in 2011, drought-stricken Oklahoma saw the hottest summer ever recorded for a U.S. state. Climatologist Kevin Trenberth quantified the impact of human caused warming this way in an email: "The extra heat from the increase in heat trapping gases in the atmosphere over six months is equivalent to running a small microwave oven at full power for about half an hour over every square foot of the land under the drought."

As one specific example, a 2014 study, "How unusual is the 2012-2014 California drought?", by researchers from Woods Hole Oceanographic Institution examined two paleoclimate reconstructions of drought and precipitation for California. They found that the soil moisture deficit in the state is truly unprecedented as measured by the Palmer Drought Severity Index (PDSI):

. . . the current event is the most severe drought in the last 1200 years, with single year (2014) and accumulated moisture deficits worse than any previous continuous span of dry years. . . . In terms of cumulative severity, it is the worst drought on record (-14.55 cumulative PDSI), more extreme than longer (4- to 9-year) droughts.

The researchers note, "The current California drought is exceptionally severe in the context of at least the last millennium and is driven by reduced though not unprecedented precipitation and record high temperatures." It is the combination of reduced precipitation and record temperatures that make this a 1-in-1200-year drought. The authors conclude that "temperature could have exacerbated the 2014 drought by approximately 36%.... These observations from the paleoclimate record suggest that high temperatures have combined with the low but not yet exceptional precipitation deficits to create the worst short-term drought of the last millennium for the state of California."

Although dramatic reductions in precipitation are the major driver of record droughts, hot weather droughts are considerably worse for humans, animals, and crops than cooler weather droughts—and not just because of the greater evaporation. As California climatologist and water expert Peter Gleick told me in 2014, when it's hotter, you also have "a greater ratio of rain-to-snow" and "faster melting of snow," both of which dramatically reduce the snowpack that is such a critical reservoir for California and the West during the dry summer months.

Many regions are already seeing (1) a larger proportion of their precipitation in the form of rain than snow and (2) earlier snowmelt. A 2011 U.S. Geological Survey study found that global warming was driving a snowpack loss in the Rocky Mountains unrivaled in 800 years, which in turn was threatening the region's water supply. That study noted, "Runoff from winter snowpack—layers of snow that accumulate at high altitude—accounts for 60 to 80% of the annual water supply for more than 70 million people living in the western United States." A 2013 U.S. Geological Survey study found that "Warmer spring temperatures since 1980 are causing an estimated 20% loss of snow cover across the Rocky Mountains of western North America."

Scientists have already observed an increase in drought intensity and/or frequency due to global warming in many parts of the world. For instance, scientists had long predicted the Mediterranean region would dry out because of global warming. That drying has been observed, and at least half of it has now been attributed directly to climate change, according to a study in the *Journal of Climate* by the U.S. National Oceanic and Atmospheric Administration, "On the Increased Frequency of Mediterranean Drought."

How does climate change affect wildfires?

Global warming makes wildfires more likely and more destructive—as many scientific studies have concluded. Why? Global warming leads to more intense droughts, hotter weather, and earlier snowmelt (hence less water available for late summer and early autumn). That means wildfires are a dangerous amplifying feedback, whereby global warming causes more wildfires, which release carbon dioxide, thereby accelerating global warming.[18]

Back in 2006, the journal *Science* published "Warming and Earlier Spring Increase Western U.S. Forest Wildfire Activity," which analyzed whether the recent soaring wildfire trend was due to a change in forest management practices or to climate change. The study, led by the Scripps Institute of Oceanography, concluded it was climate change:

> Robust statistical associations between wildfire and hydroclimate in western forests indicate that increased wildfire activity over recent decades reflects sub-regional responses to changes in climate. Historical wildfire observations exhibit an abrupt transition in the mid-1980s from a regime of infrequent large wildfires of short (average of 1 week) duration to one with

much more frequent and longer burning (5 weeks) fires. This transition was marked by a shift toward unusually warm springs, longer summer dry seasons, drier vegetation (which provoked more and longer burning large wildfires), and longer fire seasons. Reduced winter precipitation and an early spring snowmelt played a role in this shift.

That 2006 study noted global warming (from human-caused emissions of greenhouse gases such as carbon dioxide) will further accelerate all of these trends during this century. The 2007 review and assessment of the scientific literature by the Intergovernmental Panel on Climate Change acknowledged the danger:

> A warming climate encourages wildfires through a longer summer period that dries fuels, promoting easier ignition and faster spread. Westerling et al. (2006) found that, in the last three decades, the wildfire season in the western U.S. has increased by 78 days, and burn durations of fires >1000 ha have increased from 7.5 to 37.1 days, in response to a spring-summer warming of 0.87°C. Earlier spring snowmelt has led to longer growing seasons and drought, especially at higher elevations, where the increase in wildfire activity has been greatest. In the south-western U.S., fire activity is correlated with ENSO positive phases [El Niños], and higher Palmer Drought Severity Indices.

By 2050, the United States will see wildfires twice as destructive as today, and some 20 million acres a year will burn, according to a 265-page federal report authored by scientists from the U.S. Forest Service. The December 2012 report found that in places such as western Colorado, which had experienced its worst wildfire ever that year, the area burned by midcentury could jump as much as fivefold.

Many other analyses on how climate change affects fire risk have come to similar conclusions. A 2012 research report by Climate Central scientists, "The Age of Western Wildfires," found that compared to 40 years ago, the wildfire burn season is two and a half months longer. The National Research Council has projected that for every degree Celsius the Earth's temperature rises, the area burned in the western U.S. could quadruple. We are on track to warm 4°C in the coming century. These findings are also in line with the observed impacts climate change is having on wildfires, where acreage burned is already on the rise.

A July 2015 study in *Nature Communications*, "Climate-induced variations in global wildfire danger from 1979 to 2013," examined the worldwide impact climate change was having on wildfires. The researchers concluded that the length of the wildfire season had increased by almost 20%, and the global burnable area more than doubled, over that time period.

Forest Service scientists spelled out other effects climate change will have on North American forests. The Rocky Mountain forests will continue to become hotter and drier, which not only boosts wildfires but also infestations of insects such as the bark beetle, which has already devastated tens of millions of acres of U.S. and Canadian forests. The mountain pine beetle alone has already wiped out forests the size of Washington State, some 70,000 square miles of trees. Milder winters mean fewer beetle larvae die, and warmer spring and fall can double their mating season. At the same time, warming allows bark beetles to extend their ranges to higher altitudes and more northern regions. The Forest Service report notes that in some cases, it appears the pine beetle can increase the risk of forest fires. The authors explain that one beetle outbreak created a "perfect storm" in 2006 in Washington, where higher elevation lodgepole pines burned "with exceptionally high intensity." Although climate change is clearly contributing to the spread of bark beetles and the devastation they cause forests, recent studies offer differing views of whether beetle-infested trees have contributed significantly to the increase in wildfires.

How does climate change affect the chances of deluges or severe precipitation?

One of the most robust scientific findings is the direct connection between global warming and more extreme precipitation or deluges. "Basic physics tells us that a warmer atmosphere is able to hold more moisture—at a rate of approximately 7% increase per degree [Celsius] warming," as the UK Met Office's Hadley Centre explained in a 2014 report titled "Climate Risk: An Update on the Science." They add, "This is expected to lead to similar percentage increases in heavy rainfall, which has generally been borne out by models and observed changes in daily rainfall."

The U.N. Intergovernmental Panel on Climate Change concluded in its comprehensive 2013 Fifth Assessment of climate science that it is likely heavy rainfall has already begun increasing over most land areas worldwide. In the United States, scientists have already observed a sharp jump in the most intense 2-day rainstorms, the kind we used to see only once every 5 years (see Figure 2.2).

The 2014 National Climate Assessment, which is the most comprehensive analysis to date of current and future U.S. climate impacts, pointed out, "The mechanism driving these changes is well understood." The congressionally mandated report by 300 leading climate scientists and experts, which was reviewed by the National Academy of Sciences, explains: "Warmer air can contain more water vapor than cooler air. Global analyses show that the amount of water vapor in the atmosphere has in fact increased due to human-caused warming.... This extra moisture is available to storm systems, resulting in heavier rainfalls. Climate change also alters

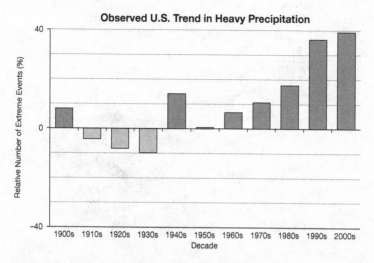

Figure 2.2 Decadal index of 2-day precipitation totals that are exceeded on average only once in a 5-year period. Changes are compared to the period 1901–1960. As data show, such once-in-5-year events have become much more common (via the 2014 U.S. National Climate Assessment).

characteristics of the atmosphere that affect weather patterns and storms."

That final point is very important. The worst deluges have jumped, but not merely because warmer air holds more moisture that in turn gets sucked into major storm systems. Increasingly, scientists have explained that climate change is altering the jet stream and weather patterns in ways that can cause storm systems to slow down or get stuck, thereby giving them more time to dump heavy precipitation (as discussed later in this chapter).

Because global warming tends to make wet areas wetter and dry areas drier, this effect does not manifest itself the same way in every part of the country. Figure 2.3 shows the 2014 National Climate Assessment chart of "percent changes in the amount of precipitation falling in very heavy events (the heaviest 1%) from 1958 to 2012 for each region".[19]

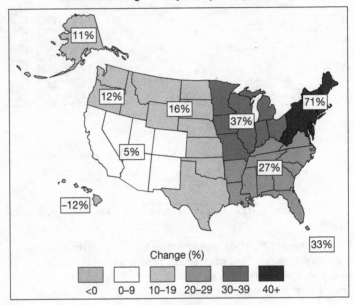

Figure 2.3 The percent changes in the amount of precipitation that fell in the heaviest events (from 1958 to 2012) for each region (via the 2014 U.S. National Climate Assessment).

Because of climate change, when it rains, it pours, literally. The 2014 National Climate Assessment explains, "The heaviest rainfall events have become heavier and more frequent, and the amount of rain falling on the heaviest rain days has also increased." Some 70% more precipitation falls in the heaviest rain events now than it did in 1958. Thus, even for the regions that are expected to see a drop in total annual precipitation (such as the Southwest), more of the precipitation they do get will be in the form of deluges so intense they can create terrible flash floods.

Finally, the UK Met Office points out that it is not merely daily rainfall extremes that are becoming more intense: "increasing evidence from observations suggests that the intensity of hourly rainfall extremes may increase more rapidly with temperature. This may be explained by latent heat released within storms invigorating vertical motion, leading to greater increases in rainfall intensity."

Does climate change mean more snow or less, worse snow storms or weaker ones?

Global warming has been observed to make the most intense rainstorms more intense. A key reason is the extra water vapor in the atmosphere from warming. This means that when it is cold enough to snow, snow storms will be fueled by more water vapor and thus be more intense themselves. Therefore, we expect fewer snowstorms in regions close to the rain-snow line, such as the central United States, although the snowstorms that do occur in those areas are likely to be more intense. It also means we expect more intense snowstorms in generally cold regions. This may appear to be counterintuitive, but the warming to date is not close to that needed to end below-freezing temperatures over large parts of the globe, although it is large enough to put measurably more water vapor into the air.[20]

In a 2014 Massachusetts Institute of Technology (MIT) study, "Snowfall in a warmer world," Professor Paul O'Gorman found that "snowfall extremes actually intensify" even many decades from now, in a future with high levels of warming. He determined "there's a narrow daily temperature range, just below the freezing point, in which extreme snow events tend to occur—a sweet spot that does not change with global warming." You may have heard of the saying, "It's too cold to snow." O'Gorman explains that it's true: "If it's very cold, there is too little water vapor in the air to support a very heavy snowfall, and if it's too warm, most of the precipitation will fall as rain."

We have long known that warmer-than-normal winters favor snowstorms. A 2006 study looked at the distribution of snowstorms from 1901-2000. It found we are seeing more northern snowstorms and that we get more snowstorms in warmer years. In mid-winter, "most of the United States had 71%–80% of their snowstorms in warmer-than-normal years." The study concluded:

> A future with wetter and warmer winters, which is one outcome expected, will bring more snowstorms than in 1901-2000. Agee (1991) found that long-term warming trends in the United States were associated with increasing cyclonic activity in North America, further indicating that a warmer future climate will generate more winter storms.

The U.S. Global Change Research Program's 2009 report, "Global Climate Change Impacts in the United States," reviewed the literature and concluded, "Cold-season storm tracks are shifting northward and the strongest storms are likely to become stronger and more frequent." So it is no surprise that a 2012 study, "When It Rains, It Pours," found extreme snowstorms and deluges are becoming more frequent and more severe.

The 2014 U.S. National Climate Assessment examined the regional trends in snow and rain in great detail. It found that, globally, "the amount of water vapor in the atmosphere has in fact increased due to human-caused warming," as scientists had predicted. That means more water vapor is available for storms of all kinds, including snowstorms when it is cold enough.

Climate change, including the loss of Arctic sea ice, can also affect entire weather patterns. The Assessment notes, "More open water can also increase snowfall over northern land areas and increase the north-south meanders of the jet stream, consistent with the occurrence of unusually cold and snowy winters at mid-latitudes in several recent years." This theory entails "significant uncertainties," and is related to the broader research into whether warming is causing changes in the atmosphere that are causing storms to get stuck, thereby giving them more time to deliver record-setting amounts of rain or snow.

The National Climate Assessment noted that this "remains an active research area" but pointed out that: "Heavier-than-normal snowfalls recently observed in the Midwest and Northeast U.S. in some years, with little snow in other years, are consistent with indications of increased blocking (a large scale pressure pattern with little or no movement) of the wintertime circulation of the Northern Hemisphere." You can see these remarkable swings in the fraction of annual precipitation coming from extreme deluges in New England from NOAA's "Climate Extremes Index" for the past century (see Figure 2.4).

A 2012 study found that Central Europe may see some colder and snowier winters in the next few decades because of climate change, particularly Arctic amplification:

> [T]he probability of cold winters with much snow in Central Europe rises when the Arctic is covered by less sea ice in summer. Scientists of the Research Unit Potsdam of the Alfred Wegener Institute for Polar and Marine Research in the Helmholtz Association have decrypted a mechanism in which a shrinking summertime sea ice cover changes the air pressure zones in the Arctic atmosphere and impacts our European winter weather.

What of the future? Dr. Kevin Trenberth explains, "In mid winter, it is expected with climate change that snowfalls will increase as long as the temperatures are cold enough, because they are warmer than they would have been and the atmosphere can hold 4% more moisture for every 1F increase in temperature. So as long as it does not warm above freezing, the result is a greater dump of snow." On the other hand, "at the beginning and end of winter, it warms enough that it is more likely for rain to result." The net result is that average total snowfall may not increase. The 2014 MIT study provides

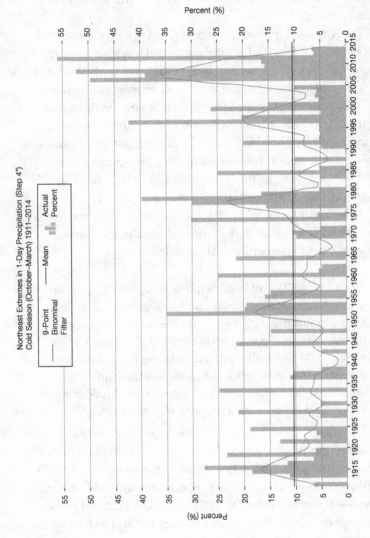

Figure 2.4 U.S. National Oceanic and Atmospheric Administration (NOAA) chart showing the percentage (×2) of New England "with a much greater than normal proportion of precipitation derived from extreme (equivalent to the highest tenth percentile) 1-day precipitation events" during the cold season (October–March).

some more regional specificity about how the future will play out in terms of average snowfall versus extreme snowfall:

> The study found that, under high warming scenarios, those low-lying regions with average winter temperatures normally just below freezing would see a 65 percent reduction in average winter snowfall. But in these places, the heaviest snowstorms on average became only 8 percent less intense. In the higher latitudes, extreme snowfall became more intense, with 10 percent more snow, even under scenarios of relatively high average warming.

How does climate change affect storm surge?

The most direct way that climate change affects storm surge is by raising sea levels. As the average sea-level rises, storm surges on average will also rise, even if the storms themselves do not become more intense. Studies also suggest that the global warming is driving an increase in the some of the most intense and damaging superstorms, which would further boost storm surge for those storms beyond the average rise in sea levels.

A 2013 study, "Hurricane Sandy Inundation Probabilities Today and Tomorrow," by NOAA researchers explored the impact of sea-level rise on storm surge. That analysis found that future Superstorm Sandy-level inundation will become commonplace in the future under business-as-usual sea-level rise projections. The U.S. National Oceanic and Atmospheric Administration notes, "The record-setting impacts of Sandy were largely attributable to the massive storm surge and resulting inundation from the onshore-directed storm path coincident with high tide." Their key finding was as follows:

> . . . climate-change related increases in sea level have nearly doubled today's annual probability of a Sandy-level flood recurrence as compared to 1950. Ongoing natural

and human-induced forcing of sea level ensures that
Sandy-level inundation events will occur more frequently in the future from storms with less intensity and
lower storm surge than Sandy.

In other words, we have nearly doubled the chances for a
Sandy-type storm surge with the several inches of sea-level
rise humans have caused to date.

Although the path and size of Sandy were driven by a combination of factors—some of which have been linked to climate change, as discussed—future Sandy-level storm surges
will result from weaker storms than Sandy as sea levels rise.
Put another way, even if we never see another storm exactly
like Sandy, Sandy-like storm surges will still become more and
more common.

The NOAA study has an "intermediate high scenario"
of 2 to 4 feet of sea-level rise by 2100 and a "high scenario"
where sea-level rises 4 to 7 feet by 2100. However, although
those labels might have been appropriate back in 2013, in 2014
we have seen multiple studies on the growing likelihood the
West Antarctic Ice Sheet faces collapse sooner than scientists
projected—and similar findings that "Greenland will be far
greater contributor to sea rise than expected."

NOAA's "intermediate high scenario" should probably be
relabeled "business as usual," and the "high scenario" renamed
the "planning case," because we typically do infrastructure
design and planning not on the best-case or most-likely case,
but rather the plausible worst-case scenario.

Even with just 2 to 4 feet of sea-level rise by 2100, the NOAA
researchers find Sandy-level storm surge events recurring
about *once a year* (or more frequently) over the vast majority of the coast from Connecticut to southern New Jersey by
century's end.

However, in the higher sea-level rise case we ought to plan
for, even areas that had the very worst storm surges from

Sandy—areas of total devastation (such as Sandy Hook and The Battery)—will be inundated by such storm surges every year or two. In fact, in that scenario, the New Jersey shore from Atlantic City south would see Sandy-level storm surges almost every year by mid-century.

Is climate change making hurricanes more destructive?

The most damaging aspect of a hurricane is the storm surge, as in the case of Hurricane Katrina and Hurricane Sandy. We have already seen that sea-level rise is increasing the chances of a Sandy-level storm surge. In addition, we know that global warming increases the intensity of rainfall from the biggest storms, which further adds to flooding. However, there is also evidence to suggest that the warming itself provides fuel for the biggest storms.[21]

Long-term tropical storm records around the world tend to be problematic because "we do a poor job estimating the intensity of storms that are not surveyed by aircraft," as Massachusetts Institute of Technology hurricane expert Dr. Kerry Emanuel explained in 2015. He notes that "Currently, only North Atlantic tropical cyclones are routinely reconnoitered by aircraft, and only if they threaten populated regions within a few days." So the best recent analyses attempt to create a consistent or homogeneous way of comparing hurricanes.

A 2012 study, led by Dr. Aslak Grinsted, created a consistent record of large storm surge events in the Atlantic over the previous nine decades. It found the worst storm surges "can be attributed to landfalling tropical cyclones"—hurricanes cause the biggest storm surges. It also found the worst storm surges "also correspond with the most economically damaging Atlantic cyclones"—hurricanes with the biggest storm surges caused the most destruction. A major finding was that Katrina-sized surges "have been twice as frequent in warm years compared with cold years."

Why does this happen? There are more active cyclones in warm years than cold years, and "The largest cyclones are most affected by warmer conditions." This is not a huge surprise given that hurricanes get their energy from warm surface waters. In fact, tropical cyclones and hurricanes are threshold events: if sea surface temperatures are below 80°F (26.5°C), they do not form. One of the ways that hurricanes are weakened is the upwelling of colder, deeper water due to the hurricane's own violent churning action. However, if the deeper water is also warm—as one would expect in warmer years—it does not weaken the hurricane. In fact, it may continue to intensify.

Typically, for a hurricane to become a Category Four or Five superstorm, it must pass over a pool of relatively deep, warm water. For instance, the National Climatic Data Center 2006 report on Katrina begins its explanation by noting that the surface temperatures in the Gulf of Mexico during the last week in August 2005 "were one to two degrees Celsius above normal, and the warm temperatures extended to a considerable depth through the upper ocean layer." The report continues, "Also, Katrina crossed the 'loop current' (belt of even warmer water), during which time explosive intensification occurred. The temperature of the ocean surface is a critical element in the formation and strength of hurricanes."

In a 2013 paper, "Projected Atlantic Hurricane Surge Threat from Rising Temperatures," Grinsted and his colleagues determined that the most extreme storm surge events "are especially sensitive to temperature changes, and we estimate a doubling of Katrina magnitude events associated with the warming over the 20th century." The study concludes, "we have probably crossed the threshold where Katrina magnitude hurricane surges are more likely caused by global warming than not."

Another 2013 paper, "Recent Intense Hurricane Response to Global Climate Change," in *Climate Dynamics*, looked at hurricane frequency and intensity in recent decades as they relate to human-emissions of greenhouse gases and aerosols. Researchers at the U.S. National Center for Atmospheric Research found no human signal in the total number of tropical cyclone or hurricane that occur each year; however, they find that "since 1975 there has been a substantial and observable regional and global increase in the proportion of Cat 4–5 hurricanes of 25–30% per °C of anthropogenic global warming."

A third 2013 paper is "Trend Analysis with a New Global Record of Tropical Cyclone Intensity," in the *Journal of Climate*. The study was led by Dr. James Kossin of NOAA's National Climatic Data Center. Hurricane expert Emanuel calls this "the best existing analysis of South Pacific tropical cyclones" in his article on Haiyan and Pam "two exceptionally intense tropical cyclones," that caused devastation in the western Pacific. The 2013 Kossin et al. paper concluded: "Dramatic changes in the frequency distribution of lifetime maximum intensity (LMI) have occurred in the North Atlantic, while smaller changes are evident in the South Pacific and South Indian Oceans, and the stronger hurricanes in all of these regions have become more intense."

Thus, the best evidence and analysis finds that although we are not seeing more hurricanes, we are seeing more of the Category 4 or 5 super-hurricanes, the ones that historically have done the most damage and that have destroyed entire coastal cities. At the same time, we are seeing a significant rise in the most damaging storm surges, whereby even a Category 1 hurricane (such as Sandy) that hits in precisely the worst possible place can cause unprecedented damage to coastal communities and major cities.

What is Arctic amplification and how does it affect extreme weather?

As climate science predicts, the Arctic is warming much faster than the rest of the globe. This is often called polar amplification. Arctic amplification accelerates the loss of land-based ice in the northern hemisphere, including the Greenland ice sheet, which accelerates sea-level rise and worsens storm surges. A number of recent studies further suggest that polar amplification is weakening the Northern hemisphere's jet stream, which in turn causes certain weather patterns including droughts, deluges, and heat waves, to get "stuck," which in turn worsens and prolongs their impact.

One piece of Arctic amplification is that global warming melts highly reflective white ice and snow, which is replaced by the dark blue sea or dark land, both of which absorb far more sunlight and hence far more solar energy. Arctic warming is amplified for other synergistic reasons, too. As the International Arctic Science Committee explained in their 2004 report, *Impacts of a Warming Arctic*:

- In the Arctic, compared with lower latitudes, "more of the extra trapped energy goes into warming rather than evaporation."
- In the Arctic, "the atmospheric layer that has to warm in order to warm the surface is shallower."
- So, when the sea ice retreats, the "solar heat absorbed by the oceans in summer is more easily transferred to the atmosphere in winter."

In addition, temperatures above thick Arctic ice can get incredibly low, but it cannot get much colder than freezing above open water.

All of this amplification leads to more snow and ice melting, further decreasing Earth's reflectivity (albedo), causing more heating, which the thinner arctic atmosphere spreads more quickly over the entire polar region, and so on and on. How much do all of the processes amplify warming in Polar Regions? A July 2010 historical study, "Arctic Amplification: Can the Past Constrain the Future?" using 3 million years of paleoclimate data found that "the Arctic temperature change consistently exceeds the Northern Hemisphere average by a factor of 3-4, suggesting that [present] Arctic warming will continue to greatly exceed the global average over the coming century, with concomitant reductions in terrestrial ice masses and, consequently, an increasing rate of sea level rise."

Another July 2010 study, "History of Sea Ice in the Arctic," by an international team of 18 scientists examined almost 300 previous and ongoing studies. The researches concluded, "The current reduction in Arctic ice cover started in the late 19th century, consistent with the rapidly warming climate, and became very pronounced over the last three decades. This ice loss appears to be unmatched over at least the last few thousand years and unexplainable by any of the known natural variabilities." In short, the near collapse in Arctic sea ice volume in recent years cannot be explained by natural variability. That same study found that "Reviewed geological data indicate that the history of Arctic sea ice is closely linked with climate changes driven primarily by greenhouse and orbital forcings and associated feedbacks. This link is reflected in the persistence of the Arctic amplification, where fast feedbacks are largely controlled by sea-ice conditions."

Therefore, external forcings—primarily orbital in the past and primarily greenhouse gases now—start a warming process that is accelerated by Arctic amplification. These July 2010 studies were part of a special-themed issue of *Quaternary Science Reviews*, "Arctic Paleoclimate Synthesis." The summary overview of that special issue makes another noteworthy point, "Taken together, the size and speed of the summer sea ice loss over the last few decades appear anomalous compared to events from previous thousands of years, especially considering that changes in the Earth's orbit over this time have made sea ice melting less, not more, likely." That is, absent human-caused global warming, orbital changes would actually be cooling the planet. A 2009 study in the journal *Science*, "Recent Warming Reverses Long-Term Arctic Cooling," found that prior to approximately 1900, the Arctic had been slowly cooling for some 2000 years, which was replaced by rapid warming only in the last century or so, driven by carbon pollution.

A 2014 study by the Scripps Institution of Oceanography in La Jolla, California analyzed just how much amplification has occurred in recent decades. Since the 1970s, the Arctic has warmed by 2°C (3.6°F). The September minimum Arctic sea ice extent dropped by 40% in the past four decades, some 35,000 square miles per year since 1979. Scripps researchers determined that between 1979 and 2011, the Arctic grew some 8% darker. The extra energy absorbed from that change is one fourth of the entire heat-trapping effect of carbon dioxide during that time. One of the study's authors explained, "Although more work is needed, a possible implication of these results is that the amplifying feedback of Arctic sea ice changes on global warming is larger than previously expected."[22]

This larger than expected Arctic amplification is a serious concern for climate scientists for several reasons. First, the faster the Arctic heats up, the faster the Greenland ice sheet melts, and the faster sea-level rise impacts coastal communities. Second, the faster the Arctic heats up, the faster the permafrost melts, and the faster it begins releasing vast amounts of heat trapping carbon dioxide and methane into the atmosphere, further accelerating global warming. Third, there is a growing body of research connecting Arctic amplification to the recent jump in extreme weather in North America. Let's look at some of that research.

Is climate change and/or Arctic amplification affecting extreme weather in the northern hemisphere?

We have seen a jump in extreme weather events in the Northern Hemisphere in the last several years. Destructive weather patterns—including droughts, deluges, and heat waves—are increasingly getting "stuck" or "blocked," which in turn worsens and prolongs their impact beyond what might be expected just from the recent human-caused increase in global temperatures.

This blocking and the unexpected jump appears to be driven in large part by a weakening of the jet stream. In addition, a growing body of research ties the recent changes in the jet stream to Arctic amplification of global warming. A 2015 study by climatologists Jennifer Francis and Stephen Vavrus, "Evidence for a Wavier Jet Stream in Response to Rapid Arctic Warming," found that Arctic amplification was leading to "more frequent high-amplitude (wavy) jet-stream configurations that favor persistent weather patterns."[23]

The jump in northern hemisphere extreme weather has been widely noted. Reinsurer Munich Re has the most comprehensive database of global natural catastrophes. Their 2010 analysis, "Large Number of Weather Extremes as Strong Indication of Climate Change," concluded, "It would seem that the only plausible explanation for the rise in weather-related catastrophes is climate change. The view that weather extremes are more frequent and intense due to global warming coincides with the current state of scientific knowledge." For instance, a 2010 *Journal of Climate* study found that "global warming is the main cause of a significant intensification in the North Atlantic Subtropical High (NASH) that in recent decades has more than doubled the frequency of abnormally wet or dry summer weather in the southeastern United States."

In 2011, Dr. Peter Höppe, Head of the Geo Risks Research Department at Munich Re, explained to me what had persuaded him of the causal link:

For me the most convincing piece of evidence that global warming has been contributing already to more and more intense weather related natural catastrophes is the fact that while we find a steep increase in the number of loss relevant weather events (about tripling in the last 30 years) we only find a slight increase in geophysical (earthquake, volcano, tsunami) events, which should not be affected by global warming. If the whole trend we find in weather related disaster should be caused by reporting bias, or socio-demographic or economic developments we would expect to find it similarly for the geophysical events.

That was before two years of off-the-charts extreme weather catastrophes, particularly in North America (discussed earlier in this chapter). In 2011, the head of NOAA said the record dozen billion-dollar weather disasters was "a harbinger of things to come."

In an October 2012 study, Munich Re linked the rapid rise in North American extreme weather catastrophes to man-made climate change: "Climate-driven changes are already evident over the last few decades for severe thunderstorms, for heavy precipitation and flash flooding, for hurricane activity, and for heat wave, drought and wild-fire dynamics in parts of North America." At the same time nonclimatic events (earthquakes, volcanoes, tsunamis) have barely changed. Höppe said at the time: "In all likelihood, we have to regard this finding as an initial climate-change footprint in our US loss data from the last four decades. Previously, there had not been such a strong chain of evidence. If the first effects of climate change are already perceptible, all alerts and measures against it have become even more pressing."

In addition, in October 2012, a study led by NOAA, "The Recent Shift in Early Summer Arctic Atmospheric Circulation," concluded that global warming was driving changes in extreme weather in North America. The lead author, NOAA's James Overland, said "Our research reveals a change in the summer Arctic wind pattern over the past six years. This shift demonstrates a physical connection between reduced Arctic sea ice in the summer, loss of Greenland ice, and potentially, weather in North American and Europe." As NOAA explained at the time:

The shift provides additional evidence that changes in the Arctic are not only directly because of global warming, as shown by warmer air and sea temperatures, but are also part of an "Arctic amplification" through which multiple Arctic-specific physical processes interact to accelerate temperature change, ice variability, and ecological impacts.

How does Arctic amplification affect the jet stream? "Enhanced warming of the Arctic affects the jet stream by slowing its west-to-east winds and by promoting larger north-south meanders in the flow," NOAA explains. "The researchers say that with more solar energy going into the Arctic Ocean because of lost ice, there is reason to expect more extreme weather events, such as heavy snowfall, heat waves, and flooding in North America and Europe but these will vary in location, intensity, and timescales."

Professor Jennifer Francis of Rutgers (coauthor of the 2012 NOAA-led study) said at the time, "What we're seeing is stark evidence that the gradual temperature increase is not the important story related to climate change; it's the rapid regional changes and increased frequency of extreme weather that global warming is causing. As the Arctic warms at twice the global rate, we expect an increased probability of extreme weather events across the temperate latitudes of the northern hemisphere, where billions of people live."

The path of the jet stream "typically has a meandering shape, and these meanders themselves propagate east, at lower speeds than that of the actual wind within the flow. Each large meander, or wave, within the jet stream is known as a Rossby wave." An August 2014 study from a team of scientists from the Potsdam Institute for Climate Impact Research provided a specific mechanism for why we are seeing this jump in extreme weather: Some Rossby waves are stalling out for extended periods of time. The study found that "in periods with extreme weather, some of these waves become virtually stalled and greatly amplified." The Potsdam Institute explained at the time:

> Weather extremes in the summer—such as the record heat wave in the United States that hit corn farmers and worsened wildfires in 2012—have reached an exceptional number in the last ten years. Man-made global warming can explain a gradual increase in periods of severe heat, but the observed change in the magnitude and duration of some events is not so easily explained. It has been linked to a recently discovered mechanism: the trapping of giant waves in the atmosphere. A new data analysis now shows that such wave-trapping events are indeed on the rise.

Not every study comes to the same conclusion as NOAA, Francis, and the Potsdam Institute, so the finding must still be considered preliminary. The interactions between climate change and Northern Hemisphere weather are complex. A 2014 *Nature Geoscience* paper, "Recent Arctic Amplification and Extreme Mid-Latitude Weather," written by 11 of the leading researchers in the field, including Francis, concludes that Arctic amplification has driven "dramatic melting of Arctic sea ice and spring snow cover, at a pace greater than that simulated by climate models"—and that this has "coincided with a period of a ostensibly more extreme weather events across the northern hemisphere mid-latitudes." The authors agree that global warming could well be driving the increase in extremes, but "large uncertainties regarding the magnitude of such of influence remain." We still have much more to learn.

That said, "Given the very large reductions in Arctic sea ice, and the heat escaping from the Arctic ocean into the overlying atmosphere, it would be surprising if the retreat in Arctic sea ice did *not* modify the large-scale circulation of the atmosphere in some way," as Michael Mann, director of the Earth System Science Center at Pennsylvania State University told me. "We now have a healthy body of research suggesting that we may indeed be already seeing this now in the form of more persistent anomalies in temperature, rainfall, and drought in North America." The 2015 study on the subject, by Francis and Vavrus, presents "new metrics and new evidence" of the link between Arctic amplification (AA) and extreme weather, concluding:

> These results reinforce the hypothesis that a rapidly warming Arctic promotes amplified jet-stream trajectories, which are known to favor persistent weather patterns and a higher likelihood of extreme weather events. Based on these results, we conclude that further strengthening and expansion of AA in all seasons, as a result of unabated increases in greenhouse gas emissions, will contribute to an increasingly wavy character in the upper-level winds, and consequently, an increase in extreme weather events that arise from prolonged atmospheric conditions.

Given how devastating many of the recent Northern Hemisphere extreme weather events have been, an increase in extreme weather is a great risk communities and countries need to start planning for seriously. This emerging area of climate science bears close watching.

Is climate change affecting tornado formation?

Tornadoes are violent and destructive weather events. Twisters have been seen on every continent (outside of Antarctica), but the overwhelming majority occur in the United States, most in the so-called Tornado Alley around Texas, Oklahoma, Kansas, and Nebraska.

April 2011 set records in the United States for most torna-does in a month and in 24 hours. The "Katrina of tornado out-breaks," as some called it, saw 758 tornadoes, with 316 on April 27 alone. NOAA notes that "The previous record for April tor-nadoes was set in April 1974 with 267 tornadoes", and "The previous record number of tornadoes during *any month* was 542 tornadoes set in May 2004." Then, in May 2011, there was another wave of tornadoes, including an usually destructive one three-fourths of a mile wide with winds in excess of 200 miles per hour, which devastated the city of Joplin Missouri, killing 157.[24]

This unusual spate of tornadoes naturally led many to ask whether there was a connection to climate change. Because this is a relatively new and rapidly evolving area of climate studies, one in which the data are difficult to analyze and the scientific literature is relatively sparse (but growing), I have consulted with leading experts on tornadoes, extreme weather, and climate change over the last few years. Tom Karl, the director of NOAA's National Climatic Data Center, explained in a 2011 e-mail: "What we can say with confidence is that heavy and extreme precipitation events often associ-ated with thunderstorms and convection are increasing and have been linked to human-induced changes in atmospheric composition."

A September 2013 study from Stanford, "Robust Increases in Severe Thunderstorm Environments in Response to Greenhouse Forcing," points to "a possible increase in the number of days supportive of tornadic storms." In particular, the *Proceedings of the National Academy of Sciences* study found that sustained global warming will boost the number of days experiencing conditions that produce severe events during spring some 40% by century's end.

Tornadoes "come from certain thunderstorms, usually super-cell thunderstorms," climatologist Dr. Kevin Trenberth has explained, but you need "a wind shear environment that promotes rotation." Global warming, it was thought, may decrease the wind shear and that may counterbalance the impact on tornado generation from the increase in thunderstorm intensity. However, the Stanford study found that most of the decline in wind shear occurred on days that were not suitable for tornado formation anyway. Trenberth, the former head of the Climate Analysis Section of the U.S. National Center for Atmospheric Research, notes:

> The main climate change connection is via the basic instability of the low level air that creates the convection and thunderstorms in the first place. Warmer and moister conditions are the key for unstable air.

> The climate change effect is probably only a 5 to 10% effect in terms of the instability and subsequent rainfall, but it translates into up to a 32% effect in terms of damage. (It is highly nonlinear). So there is a chain of events and climate change mainly affects the first link: the basic buoyancy of the air is increased. Whether that translates into a super-cell storm and one with a tornado is largely [a matter of] chance.

Many scientists would agree with the December 2013 assertion of Penn State meteorology professor Paul Markowski and National Severe Storms Laboratory senior research scientist Harold Brooks that, "Because of the inconsistency in [historical tornado] records, it is not known what effect global warming is having on tornado intensity."

At the same time, however, Florida State University researchers have recently reanalyzed the historical tornado data to make them more consistent. Lead researcher, Professor James Elsner, said in September, "The risk of violent tornadoes appears to be increasing." In particular, the trail of destruction from tornados may be getting longer and wider. As the Florida State news release noted, "The Oklahoma City tornado on May 31, 2013, was the largest tornado ever recorded, with a path of destruction measuring 2.6 miles in width." Elsner presented his findings at the annual meeting of the American Geophysical Union in December 2013:

> Beginning in 2000, tornado intensity—as measured by a twister's damage path—started rising sharply, said Elsner, of Florida State University. "I'm not saying this is climate change, but I do think there is a climate effect," he said. "I do think you can connect the dots."

In 2014, Elsner published a new study, "The increasing Efficiency of Tornado Days in the United States," in the journal *Climate Dynamics* with more findings. That study concludes: "The bottom line is that the risk of big tornado days featuring densely concentrated tornado outbreaks is on the rise. The results are broadly consistent with numerical modeling studies that project increases in convective energy within the tornado environment." Elsner told me, "I think our results suggest a climate change signal on U.S. tornado activity. Yet it is possible that at least some of the trend is due to changes in reporting practices."

Because this is a rapidly emerging area of study, it is probably best to avoid statements such as "global warming is to blame for" or "global warming caused" or "this is evidence of global warming," with regard to tornadoes. Climate Central's headline on Elsner's study offers a good way to think about the connection: "Tornado Outbreaks Could Have a Climate Change Assist." In any case, although tornadoes will continue to grab the headlines wherever they flatten cities and take lives, it is virtually certain that other extreme events—and ultimately the permanently changed climate—will cause the greatest harm attributable to human emissions of greenhouse gases.

*In a warming world, why do some winters
still seem unusually severe?*

The world's leading scientists and governments have concluded, "Warming of the climate system is unequivocal" and
a "settled fact," as discussed in Chapter One. Yet anyone who
lives in the East Coast of the United States or in various sections of Europe knows that winters seem as severe as ever. We
still have record-setting snowstorms and still set records for
cold temperatures in different places. There are several reasons why this occurs in a warming world.

First, many people, especially in the northern latitudes,
view massive snowstorms or blizzards as iconic indicators
of a severe winter. However, as seen earlier in this chapter,
warmer-than-normal winters favor snowstorms. It can be "too
cold to snow." In addition, because global warming means more
water vapor in the atmosphere available for precipitation, climate science actually projects that most extreme snowstorms
are going to get worse in a great many northern locations for the
perceivable future. That may seem counterintuitive, but that is
one goal of science, to keep our intuition from leading us astray.

Second, the world has warmed approximately 1°F since
1950. That does not mean the end of winter or the end of
record daily low temperatures. It is still going to be much,
much colder on average in January than July. As for daily temperature fluctuations, they are so large at the local level that
we will be seeing daily cold records—lowest daily minimum
temperature and lowest daily maximum temperature—for a
long, long time. That is why climatologists prefer to look at
the statistical aggregation across the country over an extended
period of time, because it gets us beyond the oft-repeated point
that you cannot pin any one single, local temperature record
on global warming.

In 2009, the U.S. National Center for Atmospheric Research
explained, "Record High Temperatures Far Outpace Record
Lows Across U.S.":

Spurred by a warming climate, daily record high temperatures occurred twice as often as record lows over the last decade across the continental United States, new research shows. The ratio of record highs to lows is likely to increase dramatically in coming decades if emissions of greenhouse gases continue to climb.

The NCAR study looked at millions of daily high and low temperature readings taken over 60 years at some 1,800 U.S. weather stations. It pointed out "If temperatures were not warming, the number of record daily highs and lows being set each year would be approximately even." Back in the 1960s and 1970s, there were actually "slightly more record daily lows than highs." However, in the 1980s, 1990s, and 2000s, "record highs have increasingly predominated, with the ratio now about two-to-one for the 48 states as a whole." Since that study, the ratio has increased further. Therefore, we can expect that for the foreseeable future, locations in the United States and around the world will still routinely see daily low temperature records set; however, overall, there will be many more high temperature records set.

Third, our perception of a cold winter is relative. As the globe warms, we will tend to think of mild winters as the norm. When we do get a cool winter, it will, relative to recent experience, seem unusually cold. This is sometimes called the phenomenon of "shifting baselines." Readers who are younger than 30 years old have never lived through a single month in which the planet's average surface temperature was below average.

3

PROJECTED CLIMATE IMPACTS

This chapter will discuss what climate science projects will happen this century. The focus will be on the so-called business-as-usual case, which assumes no significant global action to reduce greenhouse gas emissions trends in the foreseeable future. The primary source will be the latest scientific literature and interviews with leading scientists.

What kind of impacts can we expect this century from business-as-usual climate change?

The recent scientific literature warns us that we face multiple severe impacts in the coming decades if the world stays near our current greenhouse gas emissions path. These impacts include the following:

- Very high temperature rise, especially over land
- Worsening Dust Bowl conditions over the U.S. Southwest, Southern Europe, and many other regions around the globe that are heavily populated and/or heavily farmed
- Sea level rise of up to 1 foot by 2050, and 4 to 6 feet (or more) by 2100, rising as much as 12 inches or more each *decade* thereafter
- Massive species loss on land and sea
- Much more extreme weather

- Food insecurity—the increasing difficulty of feeding 7 billion, then 8 billion, and then 9 billion people in a world with an increasingly inhospitable climate
- Myriad direct and indirect health impacts

In November 2012, the World Bank issued a blunt report, "Turn Down the Heat: Why a 4°C Warmer World *Must* be Avoided." In this report, the Bank warned that "we're on track for a 4°C warmer world marked by extreme heat-waves, declining global food stocks, loss of ecosystems and biodiversity, and life-threatening sea level rise."

The April 2014 Intergovernmental Panel on Climate Change report from the world leading scientists and governments reviewing the scientific literature on climate change mitigation (greenhouse gas reduction) explained, "Baseline scenarios, those without additional mitigation, result in global mean surface temperature increases in 2100 from 3.7°C to 4.8°C compared to pre-industrial levels." Furthermore, in the baseline scenario, the world continues to warm after 2100. The fact that we are headed towards 4°C (7°F) warming or more has been clear for more than a decade in the scientific literature. However, even so, it was not until fairly recently that scientists spent much time actually exploring what such a level of warming might mean for Homo sapiens, other species, and the general livability of the planet.

The first major international scientific conference on the subject—Oxford's "4 Degrees and Beyond" conference—was not held until September 2009. In 2011, the United Kindom (UK) Royal Society devoted a special issue of one of its journals to "Four Degrees and Beyond: The Potential for a Global Temperature Increase of Four Degrees and Its Implications." The concluding piece in that special issue noted:

> . . . a 4°C world would be facing enormous adaptation challenges in the agricultural sector, with large areas of cropland becoming unsuitable for cultivation, and

declining agricultural yields. This world would also rapidly be losing its ecosystem services, owing to large losses in biodiversity, forests, coastal wetlands, mangroves and saltmarshes, and terrestrial carbon stores, supported by an acidified and potentially dysfunctional marine ecosystem. Drought and desertification would be widespread. . . .

In such a 4°C world, the limits for human adaptation are likely to be exceeded in many parts of the world, while the limits for adaptation for natural systems would largely be exceeded throughout the world.

This chapter will discuss these and other projected impacts in more detail. It will also examine the sources of uncertainty in these projections.

What are the biggest sources of uncertainty in projecting future global warming?

One of the biggest sources of confusion in the climate discussion involves the question of how much warming humanity will experience this century and what are the sources of uncertainty in that projection.

Based on our current greenhouse gas trajectory and the best estimate of how sensitive the climate is to greenhouse gas levels in the atmosphere, 4°C (7°F) total warming by 2100 (or shortly thereafter) is a reasonable projection, as noted above. There is considerable uncertainty in that number, although, unfortunately, most of the uncertainty involves the possibility of even greater warming.

The answer to the question of how much warming we will experience in the coming century (and beyond) depends primarily on four factors:

1. The so-called "equilibrium climate sensitivity"—the sensitivity of the climate to fast feedbacks such as melting

sea ice and increasing water vapor in the air. The equilibrium climate sensitivity is how much warming the Earth
would see on average if humanity were to only double
carbon dioxide levels in the air (from preindustrial levels) to 550 parts per million (ppm)—and there are no
major "slow" feedbacks. Numerous studies make clear
that the fast feedbacks, such as water vapor, are strong.

2. The actual level of carbon dioxide concentrations in the
air we hit, which, on our current emissions path, is far
beyond 550 ppm. We are currently at 400 ppm and rising
at more than 2 ppm a year—a rate of rise that has been
accelerating in recent decades. In the 1960s and 1970s, the
annual rate of rise of carbon dioxide levels was approximately half as fast as it is today (closer to 1 ppm a year).

3. The real-world slower (decade-scale) feedbacks, such as
the melting of the permafrost and the resulting release
of carbon dioxide and methane that had been frozen in
it (discussed below). At one time, these feedbacks were
thought unlikely to matter much by 2100. Now, the best
science says that they could add substantial warming
this century—as much as 1.5°F extra warming in 2100
from the thawing permafrost alone. Yet the latest climate
models do not factor in the extra warming from melting
permafrost.

4. Where you live—since people who live in the mid-
latitudes (like most Americans and most Europeans) are
projected to warm considerably more than the global
average. Therefore, if the planet as a whole warms 4°C
(7°F), much of the global population faces warming of
5°C (9°F) or more.

One of the reasons that there is some confusion in the public discussion of future warming is that many science communicators, including many in the media, focus on just no. 1,
the equilibrium or fast-feedback climate sensitivity. The U.N.
Intergovernmental Panel on Climate Change (IPCC) said in
its 2007 Fourth Assessment that the fast-feedback sensitivity

is "likely to be in the range 2 to 4.5°C with a best estimate of about 3°C, and is very unlikely to be less than 1.5°C. Values substantially higher than 4.5°C cannot be excluded, but agreement of models with observations is not as good for those values." Although the majority of studies tend to be in the middle of the range, some have been near the low end and some have been at the higher end. For the 2013 Fifth Assessment Report (AR5), the IPCC slightly changed the likely range to 1.5°C to 4.5°C.

Focusing on the fast-feedback sensitivity perhaps made sense a decade ago when there was some reasonable chance of stabilizing at 550 ppm atmospheric concentrations of carbon dioxide (double the preindustrial level) and some reasonable possibility that the slow feedbacks might not matter. The scientific community has focused on modeling a 550 ppm world and then a 450 ppm world because those targets have been the focus of international policy over the past quarter century. Because the IPCC was set up explicitly to provide the major governments of the world with the scientific basis for policy action, scientists generally expected governments to follow their advice. Governments did not. Now, however, the chances of greatly exceeding 550 ppm are substantial, much greater than 50% on our current trajectory. Indeed, the publication of the IPCC's 2007 Assessment revealed that the business-as-usual case (no climate policy) put us on track to 1000 ppm. In such a world of high greenhouse gas emissions and concentrations, it does not matter a great deal whether the fast-feedback sensitivity is 3°C, or, say, only 2.5°C. Either way, the world is going to get very, very warm.

Likewise, although it may have been reasonable to believe decades ago that the slower feedbacks would not play a major role this century in global warming, observations and analysis make that view untenable today. That is especially true because most of those feedbacks, such as the defrosting of the permafrost, are themselves temperature dependent. The hotter the Earth gets, the stronger and quicker these feedbacks will be. Therefore, the fact that we are headed towards relatively high levels of greenhouse gas concentrations greatly increases the risk posed by these amplifying feedbacks.

However, because these amplifying feedbacks have been difficult to model, they are generally ignored. If we want to understand their potential impact, we have to go back in time.

What do previous hot periods in Earth's climate tell us about what the future may hold in store?

The historical or paleoclimate record has a lot to tell us about what future warming may be like. The major climate changes of the past all occurred because the climate was driven to change by some external change, called a climate forcing, as discussed in Chapter One. These forcings includes changes in the intensity of the sun's radiation, volcanic eruptions, rapid releases of greenhouse gases, and changes in the Earth orbit. By studying the past, we can learn how sensitive the climate system is to such forcings.[25]

Although reconstructing the state of the climate thousands or millions of years ago is challenging, one benefit of using paleoclimate data is that we can glean information about the climate's sensitivity to all feedbacks, not just the fast ones. Humanity has raised atmospheric carbon dioxide levels to 400 ppm by volume (ppmv) from preindustrial levels of approximately 280 ppmv. A 2013 study in the journal *Science* study of paleoclimate temperatures—based on "the longest sediment core ever collected on land in the Arctic"—revealed what happened the last time the Earth had similar carbon dioxide levels. The lead researcher explained:

> One of our major findings is that the Arctic was very warm in the middle Pliocene and Early Pleistocene—roughly 3.6 to 2.2 million years ago—when others have suggested atmospheric carbon dioxide was not much higher than levels we see today. This could tell us where we are going in the near future. In other words, the Earth system

response to small changes in carbon dioxide is bigger
than suggested by earlier models.

How much warmer? The "first continuous, high-resolution
record of the Middle Pliocene" documented "sustained
warmth with summer temperatures of about 59 to 61 degrees
F [15 to 16 degrees C], about 8 degrees C [14 F] warmer than
today." This period of sustained Arctic warmth "coincides, in
part with a long interval of 1.2 million years when the West
Antarctic Ice sheet did not exist." In fact, sea levels during the
mid-Pliocene were 25 meters [82 feet] higher than today.

A 2009 analysis also in *Science* found a comparable result in
its examination of an even earlier time carbon dioxide levels
were at current levels, some 15 to 20 million years ago. As the
lead author explained, back then "global temperatures were 5
to 10 degrees Fahrenheit higher than they are today, the sea
level was approximately 75 to 120 feet higher than today, there
was no permanent sea ice cap in the Arctic and very little ice
on Antarctica and Greenland." That study concluded, "This
work may support a relatively high climate sensitivity."

In 2011, *Science* published a major review and analysis of
paleoclimate data, "Lessons from Earth's Past." The review
of the paleoclimate data suggested that carbon dioxide "may
have at least twice the effect on global temperatures than currently projected by computer models." It found that over the
next century, we could hit carbon dioxide levels last seen when
the Earth was 29°F (16°C) hotter than today.

The recent Paleoclimate work is in line with other many
major studies. For instance, scientists examined deep marine
sediments extracted from beneath the Arctic to understand
the Paleocene Eocene thermal maximum, a period some
55 million years ago of "widespread, extreme climatic warming that was associated with massive atmospheric greenhouse gas input." That 2006 study published in the journal
Nature found Artic temperatures as much as 23°C (41°F)
higher—temperatures that were nearly twice as warm as

current climate models had predicted when applied to this period. The three dozen authors conclude that existing climate models are missing crucial feedbacks that can significantly amplify polar warming.

How long could it take for this extra warming to show up? A 2006 study, "Missing Feedbacks, Asymmetric Uncertainties, and the Underestimation Of Future Warming," examined temperature and atmospheric changes during the past 400,000 years. It found significant increases in both carbon dioxide and methane concentrations as temperatures rise. According to this *Geophysical Research Letters* paper, if our current climate models correctly accounted for such "missing feedbacks," then "we would be predicting a significantly greater increase in global warming than is currently forecast over the next century and beyond" perhaps as much as 1.5°C warmer in this century alone. Let's look at some of these missing feedbacks.

How could the thawing permafrost speed up global warming beyond what climate models have projected?

The thawing tundra or permafrost may well be the single most important amplifying carbon-cycle feedback. Yet, none of the Intergovernmental Panel on Climate Change's climate models include carbon dioxide or methane emissions from warming tundra as a feedback. Therefore, those models likely underestimate future warming.

The tundra or permafrost is soil that stays below freezing (0°C or 32°F) for at least 2 years. Normally, plants capture carbon dioxide from the atmosphere during photosynthesis and slowly release that carbon back into the atmosphere after they die. However, the Arctic acts like a freezer—a very large carbon freezer—and the decomposition rate is very low, or at least it has been. We are in the process of leaving the freezer door wide open. The tundra is being transformed from a long-term carbon locker to a short-term carbon unlocker.

How large? The permafrost contains more than some 1.5 trillion tons of frozen carbon, which is nearly twice as much carbon as contained in the atmosphere. Although most of the carbon in defrosting permafrost would probably be released as carbon dioxide, some of it would be released as methane, a much stronger heat-trapping gas. That is especially true in areas of the tundra that are frozen wetlands or bogs. The Siberian frozen bog is estimated to contain 70 billion tons of methane. If the bogs become drier as they warm, the methane (CH_4) will oxidize and the emissions will be primarily carbon dioxide (CO_2). However, if the bogs stay wet, as many have in recent years, the methane will escape directly into the atmosphere.

Methane is 34 times more potent at trapping heat as carbon dioxide over a 100-year time horizon, but it is 86 times as potent over 20 years. Because we are worried about decade-scale feedbacks such as the permafrost, we should focus on shorter time frames than one century. Some 500 million metric tons of methane are emitted each year from natural and human sources, so if even a small fraction of the 70 billion tons of methane in the Siberian bogs escapes, it will swamp those emissions and dramatically accelerate global warming. Researchers monitoring a single Swedish bog, or mire, found it had experienced a 20% to 60% increase in methane emissions between 1970 and 2000. In some methane hotspots in eastern Siberia, "the gas was bubbling from thawing permafrost so fast it was preventing the surface from freezing, even in the midst of winter," as *New Scientist* reported in 2005.[26]

Also in 2005, a study that was led by National Center for Atmosphere Research climatologist David Lawrence found that nearly all of the top 11 feet (3.5 meters) of permafrost around the globe could disappear by the end of this century. Using the first "fully interactive climate system model" applied to study permafrost, the researchers found that if we were to stabilize carbon dioxide concentrations in the air at 550 parts per million this century, the permafrost would plummet from over 4 million square miles today to 1.5 million. Remember that the Arctic region warms up much faster than the planet as a whole. In fact, a 2008 *Geophysical Research Letters* study by leading tundra experts, "Accelerated Arctic Land Warming and Permafrost Degradation During Rapid Sea Ice Loss," concluded the following: "We find that simulated western Arctic land warming trends during rapid sea ice loss are 3.5 times greater than secular 21st century climate-change trends. The accelerated warming signal penetrates up to 1500 km inland." The last decade has been a time of very rapid sea ice loss, with a record-breaking drop in both sea ice surface area and volume. That result suggests accelerated warming 1500 kilometers inland (930 miles), which is precisely where the tundra is found.

Scientists have continued to refine estimates of permafrost loss from various emissions scenarios. A 2011 study by the U.S. National Oceanic and Atmospheric Administration (NOAA) and the National Snow and Ice Data Center found that thawing permafrost will turn the Arctic from a place that stores carbon (a sink) to a place that generates carbon (a source) in the 2020s—and release a hundred billion tons of carbon by 2100. That study, "Amount and Timing of Permafrost Carbon Release in Response To Climate Warming," concluded:

> The thaw and release of carbon currently frozen in permafrost will increase atmospheric CO_2 concentrations and amplify surface warming to initiate a positive permafrost carbon feedback (PCF) on climate. . . . We predict that the PCF will change the arctic from a carbon sink to a source after the mid-2020s. . . . The thaw and decay of permafrost carbon is irreversible and accounting for the PCF will require larger reductions in fossil fuel emissions to reach a target atmospheric CO_2 concentration.

The study acknowledges that it almost certainly underestimates the warming the permafrost carbon feedback will cause.

It assumes all of the carbon released will come out as carbon dioxide, with no methane. It does not attempt to calculate and incorporate the extra warming from that carbon dioxide. In addition, it assumes we get on a path of human-caused greenhouse gases that is lower than our current path and results in relatively moderate warming. Even so, "The amount of carbon released [by 2200] is equivalent to half the amount of carbon that has been released into the atmosphere since the dawn of the industrial age," says lead author Dr. Kevin Schaefer. "That is a lot of carbon."

A December 2011 article in *Nature*, "Climate Change: High Risk of Permafrost Thaw," surveyed 41 international experts, called the Permafrost Carbon Network, who publish on permafrost issues. They concluded, "Our collective estimate is that carbon will be released more quickly than models suggest, and at levels that are cause for serious concern." They project as much as 380 billion metric tons of carbon dioxide equivalent will be released by 2100. They calculate that the defrosting permafrost will release roughly the same amount of carbon as current rates of deforestation would release. However, "because these emissions include significant quantities of methane, the overall effect on climate could be 2.5 times larger."

In October 2012, a *Nature Geoscience* study, "Significant Contribution to Climate Warming from the Permafrost Carbon Feedback," calculated that on our current business-as-usual greenhouse gas emissions path, the melting permafrost would add 100 ppm in additional carbon dioxide. They project that over a range of scenarios, including ones where we take strong action to reduce greenhouse gas emissions, by 2100 the defrosting permafrost would add 0.25°C (0.4°F) and possibly as much as 0.8°C (nearly 1.5°F). The scientists further note that in the business-as-usual emissions case, it is entirely possible that the rate of release of carbon from the permafrost post-2100 becomes larger than the rate at which the oceans can absorb carbon dioxide. In that case, "CO_2 will continue to build up in the atmosphere, further warming the surface and driving a self-sustaining carbon-cycle feedback."

In April 2015, the journal *Nature* published, "Climate Change and the Permafrost Carbon Feedback," the most comprehensive study of the permafrost carbon feedback to date. Researchers concluded the thawing permafrost would release some 90 billion metric tons of carbon to the atmosphere by 2100 on our current emissions path, bumping up CO_2 concentrations some 60 to 80 ppm. Emissions from the permafrost would continue at a rapid rate into the next century and beyond.

Given that the scientific literature makes clear how significant the permafrost thawing could be to projected warming this century and beyond, it was particularly surprising in December 2012 when the United Nations Environment Programme study "Policy Implications of Warming Permafrost" reported this news:

> The effect of the permafrost carbon feedback on climate has not been included in the IPCC Assessment Reports. None of the climate projections in the IPCC Fourth Assessment Report include the permafrost carbon feedback (IPCC 2007). Participating modeling teams have completed their climate projections in support of the Fifth Assessment Report, but these projections do not include the permafrost carbon feedback. Consequently, the IPCC Fifth Assessment Report, due for release in stages between September 2013 and October 2014, will not include the potential effects of the permafrost carbon feedback on global climate.

In short, any time this book or any news report cites an IPCC projection of future warming or future climate impacts, it is almost certain that projection represents an underestimate of what is to come.

How could an increase in wildfires speed up global warming beyond what climate models have projected?

Global warming increases the conditions for wildfires in many regions, as we have discussed. Trees and vegetation convert carbon dioxide into oxygen, acting as a sink or storage media for carbon. When trees and vegetation are killed, their ability to absorb carbon dioxide from the atmosphere ends. In addition, because they store a considerable amount of carbon, when they burn, the biomass gets converted back into carbon dioxide, which is then released into the atmosphere. That process causes further warming, which in turn worsens wildfires. That scenario is a classic positive, or amplifying, feedback.[27]

Consider the northern "boreal" (subarctic) forests of Canada, Russia, and Alaska. A 2013 study concluded, "Recent burning of boreal forests exceeds fire regime limits of the past 10,000 years." There are now 20,000 wildfires per 1,000 years—double the rate of 500 to 1000 years ago. The lead author of that study explained to *LiveScience*, "There's a pretty clear link between humans inducing a warmer climate and increased forest burning." Boreal forests store more than 30% of all the carbon stored on land (in vegetation and soil). Although tropical forests get most of the attention, they store a little more than half the carbon per acre that boreal forests do.

To make matters worse, much of the boreal forests rest on permafrost and peatland—both of which release a massive amount of carbon dioxide when burned. In addition, the fires blacken the area above permafrost, which then absorbs more heat from the sun, further speeding up the defrosting of the permafrost and the loss of soil carbon to the atmosphere. Most of the world's wetlands are peat, which are better known as bogs, moors, mires, and swamp forests. Peat is one of the earliest stages in the long-term processes of forming coal. It burns easily and in fact is used widely for fuel.

"Smouldering peat fires already are the largest fires on Earth in terms of their carbon footprint," explained mega-fire expert Professor Guillermo Rein. He is coauthor of a 2015 study in *Nature Geoscience*, "Global Vulnerability of Peatlands to Fire and Carbon Loss," which warns that massive, difficult to stop peatland fires are likely to become even larger in the future, since human activity keeps drying them out. Because a key reason many peatlands will become drier is global warming, and because peatland fires can release staggering amounts of carbon dioxide, this process is a vicious circle, a dangerous amplifying carbon cycle feedback. The study explains why the loss of peatlands is of such great concern to scientists: "Globally, the amount of carbon stored in peats exceeds that stored in vegetation and is similar in size to the current atmospheric carbon pool."

Massive Indonesia peatland fires during the hot and dry El Niño of 1997 and 1998 burned almost 25 million acres, among the largest set of forest fires in the past 200 years. A 2002 *Nature* analysis estimated the CO_2 released by those fires was "equivalent to 13–40% of the mean annual global carbon emissions from fossil fuels, and contributed greatly to the largest annual increase in atmospheric CO_2 concentration detected since records began in 1957."

Why do the peat fires release so much carbon? As one soil scientist explained in a November 2014 essay, it is typical in Indonesia that "Even after the forest fires end, the peat continues to smolder underground until all organic matter has completely burned into ashes." A 2008 *Nature Geoscience* study, "High Sensitivity of Peat Decomposition to Climate Change Through Water-Table Feedback," projected that "a warming of 4°C causes a 40% loss of soil organic carbon from the shallow peat and 86% from the deep peat" of Northern peatlands. On our current emissions path, the world is set to warm well beyond 4°C (7°F). According to the 2008 study, "We conclude that peatlands will quickly respond to the expected warming in this century by losing labile soil organic carbon during dry periods."

The National Aeronautics and Space Administration (NASA) explained why forests and bog land in Siberia had been burning for months in mid-2012: "Contributing to the record fires have been the record temperatures of this past summer." That summer, Siberia saw average temperatures of 93°F, which are not exactly the temperatures anyone associates with Siberia. The result, as NASA explained: "the fires burning in Russia will have worldwide effects as the torched peat bogs whose layers consist of dead plant materials will end up releasing large quantities of carbon dioxide into the air accelerating the greenhouse effect and making the air nearly unbreathable."

A 2011 study led by University of Guelph professor Merritt Turetsky found that "drying of northern wetlands has led to much more severe peatland wildfires and nine times as much carbon released into the atmosphere." Turetsky noted at the time, "Our study shows that when disturbance lowers the water table, that resistance disappears and peat becomes very flammable and vulnerable to deep burning." And that is when peatlands turn from a CO_2 sink to a CO_2 source. Turetsky also led the 2015 peatlands study in *Nature Geoscience*. It explains that "drying as a result of climate change and human activity lowers the water table in peatlands and increases the frequency and extent of peat fires." Tragically, Indonesia has drained a great deal of its peatlands—and even burned forested areas—to create palm oil plantations, a key reason there are so many forest fires and smoldering peat fires, which often ruin the air quality in the region.

"The scary thing is future climate change may actually do the same thing: dry out peatlands," explained another coauthor, climatologist Guido van der Werf. "If peatlands become more vulnerable to fire worldwide, this will exacerbate climate change in an unending loop." As several 2014 studies made clear, climate change will dry out and Dust-Bowlify large parts of the planet's arable landmass. The 2015 study concludes that "almost all peat-rich regions will become more susceptible to drying and burning with a changing climate."

What are some other key positive or amplifying feedbacks affecting the climate system?

A central worry of climate scientists is that because of climate change, the ocean and land sinks, which currently absorb more than half of all total human-caused carbon dioxide emissions, will become increasingly ineffective at absorbing carbon dioxide. That would mean a greater and greater fraction of carbon pollution would stay in the air, which would speed up climate change, causing more carbon dioxide to stay in the air—an amplifying feedback. For instance, as noted, as global warming increases forest fires and peatland fires, burning trees and vegetation, that turns one part of the land carbon "sink" will into a "source" of atmospheric carbon dioxide. Likewise, the defrosting of the permafrost and the resultant release of carbon dioxide and methane also turns part of the land sink into a source of airborne carbon dioxide.

Do we have any evidence that the land and ocean sinks are becoming less effective at taking carbon dioxide out of the atmosphere? In September 2014, the World Meteorological Organization (WMO) announced:

> The observations from WMO's Global Atmosphere Watch (GAW) network showed that CO_2 levels increased more between 2012 and 2013 than during any other year since 1984. Preliminary data indicated that this was possibly related to reduced CO_2 uptake by the Earth's biosphere in addition to the steadily increasing CO_2 emissions.[28]

Two months earlier, a similar conclusion was reached by a more comprehensive international study, "The declining uptake rate of atmospheric CO_2 by land and ocean sinks." That study, published in *Biogeosciences*, noted that, for the last five decades, 44% of total human-caused carbon dioxide emissions stayed in the atmosphere. It defined the "The CO_2 uptake rate by land and ocean sinks" as "the combined land–ocean CO_2 sink flux per unit mass of excess atmospheric CO_2 above pre-industrial concentrations." This is a measure of the land and ocean "sink efficiency," and they label it "the CO_2 sink rate." The study found that this sink rate "declined over 1959–2012 by a factor of about 1/3, implying that CO_2 sinks increased more slowly than excess CO_2."

What does declining sink efficiency mean in simple terms? "For every ton of carbon dioxide we emit into the atmosphere, we are leaving more and more in the atmosphere" each passing year, as study coauthor Josep Canadell explained to me. Canadell is the executive director of the Global Carbon Project, a group of the world's leading experts on the global carbon cycle, formed "to assist the international science community to establish a common, mutually agreed knowledge base supporting policy debate and action."

Significantly, the study found that of the reasons for the decline in land and ocean sink efficiency, "intrinsic" carbon-cycle feedbacks were responsible for 40% of the drop:

Fifth, our model-based attribution suggests that the effects of intrinsic mechanisms (carbon-cycle responses to CO_2 and carbon–climate coupling) are already evident in the carbon cycle, together accounting for ~40 % of the observed decline in [the sink rate] over 1959–2013. These intrinsic mechanisms encapsulate the vulnerability of the carbon cycle to reinforcing system feedbacks. . . . An important open question is how rapidly the intrinsic mechanisms and associated feedbacks will contribute to further decline in [the sink rate] under various emission scenarios.

The study notes, "Many (though not all) of these [feedbacks] are fundamentally nonlinear." It concludes that "Using a carbon–climate model continuing future decreases in [the sink rate] will occur under all plausible CO_2 emission scenarios." So the land and ocean sinks are projected to become increasingly less efficient, with uncertainty about exactly how fast that will happen, but a real possibility it will happen faster than it has.

We have already seen some feedbacks (such as the permafrost melt and wildfires) that reduce the net uptake of carbon dioxide from the land sink. A 2012 study led by the UK Met Office's Hadley Centre, "High Sensitivity of Future Global Warming to Land Carbon Cycle Processes," used a major global climate model to systematically study potential land carbon-cycle feedbacks. The researchers found that those feedbacks were "significantly larger than previously estimated." Those feedbacks are so large that they could add as much as a few hundred parts per million to carbon dioxide levels in 2100 compared with the no-land-feedback case, even in a scenario of moderate carbon dioxide emissions. That in turn could add 1°C or more to total warming in that case, and that is just for this century.

The oceans similarly have feedback processes that threaten to reduce their net uptake of carbon dioxide over time. For instance, global warming drives ocean stratification—the separation of the ocean into relatively distinct layers—which in turn reduces the ability of the oceans to take up carbon dioxide. Here is why. Over 90% of the heat from human-caused global warming ends up in the ocean. Most of this additional heat has ended up in near surface waters, causing those waters to warm some 0.7°C in the last century. The deeper ocean has warmed far less. However, warmer seawater is less dense, and cold seawater is denser. The result, as a 2011 article published by the UK Royal Society explained, is as follows: "This differential heating of the water column has increased the density gradient between the near-surface waters and the deep ocean, increasing the upper ocean stratification." In other words, the already less dense surface waters become even less dense. This stratification "tends to decrease upper ocean mixing and transport, thereby more strongly separating the upper ocean, which is in ready exchange with the atmosphere from the intermediate and deep ocean."

Ocean warming and the resulting stratification limit ocean uptake of carbon dioxide and other key greenhouse gases, "as the transport of these gases from the near-surface ocean into the ocean's interior is the primary rate-limiting step." How limited? Models suggest that ocean uptake of atmospheric CO_2 could be reduced by 14 to 67 billon metric tons of carbon a year—for each degree Celsius of surface warming. That could mean a drop in CO_2 uptake by 2100 of as much as 30%, which would be a major amplifying feedback because it would weaken the ability of the ocean to act as a carbon dioxide sink.

At the same time, ocean acidification itself may speed up total warming this century as much as 0.9°F, according to a 2013 study. Researchers at Germany's Max Planck Institute for Meteorology have found "Global warming amplified by reduced sulphur fluxes as a result of ocean acidification," as they titled their *Nature Climate Change* study. Sulphur in the air comes mainly from the ocean and helps form clouds that keep the Earth cool. As the journal *Nature* explained, "Phytoplankton—photosynthetic microbes that drift in sunlit water—produces a compound called dimethylsulphide (DMS). Some of this enters the atmosphere and reacts to make sulphuric acid, which clumps into aerosols, or microscopic airborne particles. Aerosols seed the formation of clouds, which help cool the Earth by reflecting sunlight." However, as the ocean acidifies, seawater appears to generate less DMS. If DMS dropped globally because of ocean acidification, it would create an amplifying feedback that would boost global warming beyond what the climate models are predicting.

How much extra warming could occur because of ocean acidification? The Max Planck Institute found that reductions in DMS would increase temperatures up to 0.48 K (0.9°F). They concluded, "Our results indicate that ocean acidification has the potential to exacerbate anthropogenic warming through a mechanism that is not considered at present in projections of future climate change." Recall that the carbon feedback from the thawing permafrost—also unmodeled by the Intergovernmental Panel on Climate in its latest assessment of the science—could add up to 1.5°F to total global warming by 2100. That means actual warming this century could be 2°F higher than the IPCC projects.

What will the impacts of sea-level rise be?

What science has told us in 2014 and 2015 about likely sea level rise this century and beyond has been shocking. Whereas many of the leading sea level rise experts had previously told me the likely range for the total rise by 2100 was 2 to 6 feet, now leading researchers believe the low range is increasingly unlikely, and the top range—the plausible worst-case scenario—is considerably above 6 feet. At the same time, the projection of how fast sea levels will be rising by century's end has also increased, with one leading expert putting the number at 1 foot per decade.

The Intergovernmental Panel on Climate Change, in its 2013 review of the scientific literature, had projected that sea-level rise by 2100 would be 0.52 to 0.98 meters (20 to 39 inches) in the business-as-usual case. The rate of sea-level rise during 2081 to 2100 could hit 1.6 centimeters a year (6 inches a decade). In that estimate, the IPCC assumed most sea-level rise will come from thermal expansion of the ocean and melting of inland glaciers around the world. They assumed the Greenland Ice Sheet makes at most a modest contribution and the Antarctic Ice Sheet makes a small net contribution. The panel acknowledges, however, that it really has no idea what the ice sheets could do: "The basis for higher projections of global mean sea-level rise in the 21st century has been considered and it has been concluded that there is currently insufficient evidence to evaluate the probability of specific levels above the assessed likely range."

To get a better projection, a 2014 study replaced "replaced the AR5 projection uncertainties for both ice sheets with probability distribution function calculated from the collective view of thirteen ice sheet experts" determined in a January 2013 study. That study concluded, "seas will likely rise around 80 cm" [31 inches] by 2100, and that "the worst case [only a 5% chance] is an increase of 180 cm [6 feet]." The ice sheet experts back then had a similar view of the most likely contribution from both ice sheets, but a much higher worst-case estimate for both, especially Antarctica.[29]

The authors of the 2014 study explain, "We acknowledge that this may have changed since its publication. For example, it is quite possible that the recent series of studies of the Amundsen Sea Sector and West Antarctic ice sheet collapse will alter expert opinion." In fact, these recent studies have altered the opinion of many experts I have spoken to.

In May 2014, we learned that the West Antarctic Ice Sheet (WAIS) appears close to if not past the point of irreversible collapse. That same month, we also learned that "Greenland's icy reaches are far more vulnerable to warm ocean waters from climate change than had been thought." We learned in August that Greenland and WAIS more than doubled their rate of ice loss in the last 5 years. Both Greenland and WAIS each have enough ice to raise sea levels some 15 to 20 feet. One of the authors of the WAIS collapse work, NASA's Eric Rignot, told me at the time, "I think that the *minimum* will be the upper end of the IPCC projections (90 cm [3 feet]) by 2100."

In 2015, researchers reported more remarkable results. First, a large glacier in the East Antarctic Ice Sheet turns out to be as unstable and as vulnerable to melting from underneath as WAIS is. This alone could "could lead to an extreme thaw increasing sea levels by about 11.5 feet (3.5 meters) worldwide if the glacier vanishes." Second, two new studies find that global warming is weakening a crucial ocean circulation in the North Atlantic, the Gulf Stream system, to a level "apparently unique in the last thousand years." In addition, if that circulation continues to weaken, it would also add another few feet of sea-level rise to the U.S. East Coast. Indeed, this weakening maybe one reason why large parts of the East Coast are already experiencing much faster sea-level rise than the rest of the world. Another 2015 study found that global sea-level rise since 1990 has been speeding up even faster than we knew. "The sea-level acceleration over the past century has been greater than had been estimated by others," explained lead author Eric Morrow. "It's a larger problem than we initially thought."

The recent findings have led top climatologists to conclude that we are likely headed toward what used to be the high-end of projected global sea-level rise this century (i.e., 4 to 5 feet) and that the worst-case scenarios where humanity fails to take aggressive action to cut greenhouse gas emissions are considerably higher than that. Such sea level rise would have severe direct consequences coastal populations. One 2015 study found that by 2060, population in the low-elevation coastal zone—below 10 meters above sea level—could hit 1.3 billion people, which is twice current levels. With even a relatively modest sea-level rise of 21 centimeters (8 inches) by then, more than 400 million people could be in the flood plain of the once-in-a-100-year storm surge. It is clear that sea-level rise post-2050 could be considerably higher than that, upwards of 10 feet (3 meters) by 2100.

On top of that, we have the danger posed to coastal areas from worsening storm surge, discussed in Chapter Two. With the kind of sea level rise many scientists are now expecting, Superstorm Sandy-level storm surges would become commonplace on the East Coast by mid-century.

So it seems very likely that hundreds of millions of people will need to relocate this century just from sea-level rise and threat of storm surge alone. Certainly, many countries will pursue a strategy of trying to adapt, building sea walls and the like. However, seawalls are very expensive, even more so if you try to design them to deal with the 100-year storm surge, and most major countries—including the United States—could not afford to protect its entire coast. In addition, some coastal geologies simply cannot be protected by sea walls, such as South Florida's.

Also, on our current emissions path, we could be seeing sea-level rise of 1 foot per decade by century's end, according to Harold Wanless, chair of University of Miami's geological sciences department. How exactly do we adapt to that? Wanless told National Geographic in 2013, "I cannot envision southeastern Florida having many people at the end of this century." In 2014, he said, "Miami, as we know it today, is doomed. It's not a question of if. It's a question of when."

One final serious impact of sea-level rise is increasing saltwater intrusion in coastal agricultural areas. As the salty ocean waters rise, they increase the solidity of coastal waters and soils. A 2015 study on salinization of coastal Bangladesh found "climate change will cause significant changes in river salinity in the southwest coastal region during the dry season (October to May) by 2050. This will likely lead to shortages of drinking and irrigation water and cause changes in aquatic ecosystems." At the same time, the rise in soil salinity could reduce yields "by 15.6% of high-yielding-variety rice and reduce the income of farmers significantly in coastal area." The study concluded "households in areas with high inundation and salinization threats have significantly higher out-migration rates for working-age male adults, dependency ratios and poverty incidence than households in non-threatened areas."

Bangladesh is not alone. Saltwater intrusion is affecting coastal ground water supplies and soils around the world. Warming-driven sea-level rise is already leading to more and more saltwater intrusion into the agriculturally rich Nile Delta, which is a cornerstone of the food supply for 80 million Egyptians. Mahmoud Medany, a researcher at Egypt's Agricultural Research Center, spoke to the *New York Times* in 2013 about the intrusion. "The Nile is the artery of life, and the Delta is our breadbasket," he said. "And if you take that away there is no Egypt."

How will climate change lead to more destructive superstorms this century?

We have already seen a few ways climate change will make superstorms more destructive, including higher storm surge and greater precipitation. Climatologist Kevin Trenberth has explained, "For a one degree Fahrenheit increase in air temperature the water holding capacity goes up by four percent." So a 5°F increase means an approximately 20% increase in atmospheric water vapor. We could see as much as 10°F warming in the coming century, which would mean a 40% increase. How will this translate into worse superstorms? Scientists believe "the intervals between storms will be longer, but then when you do have the storms they are apt to be a doozie. When it rains it pours, so to speak."

Considerable research also indicates that we are seeing more atmospheric blocking patterns that can keep major weather patterns stuck for extended periods of time. This may be related to a weakening of the jet stream and possibly the accelerated warming of the Arctic. We know that when a major rain or snow system gets stuck or slows down, then the amount of precipitation falling in a region can jump to extraordinary levels, with a year's worth of rain falling in a couple of days. Climate scientists project that the rate of global warming, and the rate of Arctic warming, is poised to jump in the coming decade. If that causes more weather patterns to get stuck, then we can expect more intense deluges from slow-moving superstorms.

In addition, all future coastal superstorms will be operating in an environment with increasingly higher sea levels. This means storm surges will get worse and worse. Put another way, a much weaker storm than Hurricane Sandy will cause comparable damage when sea levels are a few feet higher. The kind of sea-level rise we appear to be headed toward will make Sandy-type storm surges commonplace events on the East Coast after mid-century.

A 2013 study by the Niels Bohr Institute found that previous research revealed "an increasing tendency for storm hurricane surges when the climate was warmer." The *Proceedings of the National Academy of Sciences* paper was aimed at calculating the "Projected Atlantic Hurricane Surge Threat from Rising Temperatures." In particular, researchers were trying to answer the question of how global warming will affect the frequency of "extreme storm surges like that from Hurricane Katrina," which devastated New Orleans and the Gulf Coast in 2005. They found that "The most extreme events are especially sensitive to temperature changes, and we estimate a doubling of Katrina magnitude events associated with the warming over the 20th century." Since 1923, we have experienced a storm surge like Katrina's every 20 years. The study found that a degree centigrade rise in global temperatures would increase the frequency of Katrina magnitude storm surges by 3 to 4 times. A 2°C warming would mean 10 times more extreme storm surges—one occurring every other year.

In Chapter Two, I reviewed a number of studies that have found the strongest and most destructive hurricanes in many regions of the world, including the North Atlantic and South Pacific, have become more intense. A 2014 study, "Recent Intense Hurricane Response to Global Climate Change," also found that "since 1975 there has been a substantial and observable regional and global increase in the proportion of Cat 4–5 hurricanes" of some 25% to 30% per degree Celsius of human-caused global warming. That study in *Climate Dynamics* projected the trend would continue to increase until category four and five hurricanes represent 40% to 50% of all hurricanes.

The future of hurricanes remains the most difficult to predict. Global warming means warmer surface water and warmer deep water. All things being equal, that would mean future hurricanes traveling the same path are going to stay stronger longer and possibly even intensify where earlier hurricanes had weakened. What we do not know is whether, in fact, all things will be equal. Perhaps global warming will create other conditions that might serve to weaken hurricanes or change their storm path. As one 2013 study put it, "global warming may also increase vertical wind shear, which is unfavorable for cyclones, although some studies find this is a minor effect."

As of 2015, we have confidence that future hurricanes will be more destructive because they will generate more precipitation and their storm surge will be augmented by rising sea levels. The literature suggests that although we may not see more hurricanes in future years, the most intense ones will get even more intense.

What kind of droughts can we expect this century?

As much as one third of the Earth's currently habited and arable land faces a near-permanent drying this century, according to several recent studies. I called this prolonged, multidecadal warming and drying, "Dust-Bowlification," in a 2011 *Nature* article, "The Next Dust Bowl," which reviewed the literature. I used that term simply because the 1930s Dust Bowl seems to be the best analogy to what is coming. However, in fact, the coming multidecadal megadroughts will be *much* worse than the Dust Bowl of the 1930—"worse than anything seen during the last 2000 years," as a major 2014 Cornell-led study put it. They will be the kind of megadroughts that in the past destroyed entire civilizations.[30]

In that 2014 *Journal of Climate* study, "Assessing the Risk of Persistent Drought Using Climate Model Simulations and Paleoclimate Data," scientists quantified the risk of devastating, prolonged drought in the southwestern U.S. and the world due to global warming. Researchers from Cornell, University of Arizona, and the U.S. Geological Survey concluded "the risk of a decade-scale megadrought in the coming century [in the Southwest] is at least 80%, and may be higher than 90% in certain areas."

The risks of a devastating U.S. megadrought this century are quite substantial in the scenario where we keep doing little or nothing to slash carbon pollution. Furthermore, the authors point out:

> We extend our analysis of megadrought risk in the western US to the rest of the world by examining raw [model] estimates of decadal drought and multi-decadal megadrought from the three RCP [emissions] scenarios. Risks throughout the subtropics appear as high or higher than our estimates for the US Southwest (e.g., in the Mediterranean, western and southern Africa, Australia, and much of South America).

This finding is consistent with a great deal of recent research. For instance, a 2012 study from the National Center for Atmospheric Research "strengthened the case" that, unless we reverse emissions trends soon, we risk having a situation by the 2060s where large swaths of the United States, Brazil, Africa, the Mideast, Australia, Southeast Asia, and Europe are routinely in severe drought. By the 2090s, "most of southern Europe and about half of the United States is gripped by extreme drought" a great deal of the time.

The actual climate these regions face in a business-as-usual future is actually worse than the 2014 study finds because the authors "based our analysis on precipitation projections." They do not look at the impact of "increases in temperatures," which worsen any drought and lead to more evaporation of surface moisture. The authors note that other studies have looked at "precipitation minus evapotranspiration," which is the overall impact on soil moisture:

> Such studies have found that drought conditions like the Dust Bowl will become normal in the Southwest and in other subtropical dry zones. If such transitions are indeed "imminent," as stated in those studies, then the risk of decadal drought is 100 percent, and the risk

of longer-lived events is probably also extremely high. By orienting our analysis around precipitation, the risks of prolonged drought we show here are in fact the lowest levels consistent with model simulations of future climates.

This is a key point. Drought can come about for two reasons, lower precipitation or higher temperatures over an extended period of time. If a region gets hit by both of those, it will suffer an unusually extreme drought, such as we have seen in California in the last few years, or Australia in the previous decade. As I discussed in *Nature*: "Warming causes greater evaporation and, once the ground is dry, the Sun's energy goes into baking the soil, leading to a further increase in air temperature." That is why the United States saw so many temperature records in the 1930's Dust Bowl. It is why in 2011, drought-stricken Oklahoma saw the hottest summer ever recorded for a U.S. state. It is why in 2014, drought-stricken California saw its hottest year on record.

Some recent studies have attempted to incorporate the impact of warming and drying on future climate. Consider the 2014 study, "Global warming and 21st century drying" in *Climate Dynamics* led by Dr. Benjamin Cook. Columbia University explains the conclusions: "the study estimates that 12% of land will be subject to drought by 2100 through rainfall changes alone; but the drying will spread to 30% of land if higher evaporation rates from the added energy and humidity in the atmosphere are considered."

The authors explain that "even regions expected to get more rainfall will see an increased risk of drought from the hotter temperatures." Columbia notes, "Much of the concern about future drought under global warming has focused on rainfall projections, but higher evaporation rates may also play an important role as warmer temperatures wring more moisture from the soil, even in some places where rainfall is forecasted to increase." This study is "one of the first to use the

latest climate simulations to model the effects of both changing rainfall and evaporation rates on future drought." It finds
"increased evaporative drying will probably tip marginally
wet regions at mid-latitudes like the U.S. Great Plains and a
swath of southeastern China into aridity."

One of first studies that made use of early models to look
at this issue was the 1990 *Journal of Geophysical Research* study,
"Potential Evapotranspiration and the Likelihood of Future
Drought," from NASA. It projected that severe to extreme
drought in the United States, then occurring every 20 years
or so, could become an every-other-year phenomenon by
mid-century.

The study by Dr. Cook and his colleagues examined what
the "new normal" would be around the world for the period
2080–2099—in the business-as-usual warming scenario (i.e.,
humanity keeps doing very little to combat climate change).
The new normal climate of Southern Europe will be extreme
drought, much worse than the U.S. Dust Bowl—hotter, drier,
and permanent. All of Europe will be transformed in ways
that are difficult to imagine today. The same new normal will
be found over Iraq and Syria and much of the Mideast.

The same new normal will be found in the breadbasket
of China and some of the most heavily populated areas of
Australia, Africa, and South America. The normal climate
of the Amazon will be much, much drier than it has been.
A 2013 study found the Amazon's dry season already lasts 3
weeks longer than it did 30 years ago. This alone makes the
lush southern Amazon more susceptible to dieback—from
both lack of rain and increased risk of wildfires. The projected
post-2050 climate for the Amazon would make dieback all
but unstoppable, which would release large amounts of CO_2
stored there into the atmosphere, an accelerating feedback for
climate change.

Finally, if this business-as-usual projection comes to pass,
the U.S. Southwest and Central Plains breadbasket enter a new
climate where the "normal" is moderate drought and in some

places even severe drought. Decades comparable to the Dust Bowl (but much hotter) will happen routinely. The new normal for the soil of most of the rest of the country will not be much moister.

This forecast for America was confirmed by a 2015 study led by NASA, "Unprecedented 21st Century Drought Risk in the American Southwest and Central Plains." As NASA explained, this new study concludes that "carbon emissions could dramatically increase risk of U.S. megadroughts," which are droughts lasting more than three decades. "Droughts in the U.S. Southwest and Central Plains during the last half of this century could be drier and longer than drought conditions seen in those regions in the last 1,000 years"

Here is what the NASA study projects could happen for all of North America if we stay on our current, business-as-usual emissions path (see Figure 3.1). The darkest areas have soil moisture comparable to that seen during the Dust Bowl.

"Droughts like the 1930s Dust Bowl and the current drought in the Southwest have historically lasted maybe a decade or a little less," explained Ben Cook, NASA climatologist and lead author. "What these results are saying is we're going to get a drought similar to those events, but it is probably going to last at least 30 to 35 years."

This study went back to the medieval Southwest megadroughts for comparison. How bad were they? Lisa Graumlich, Dean of the University of Washington's College of the Environment, explained that the Southwest drought from 1100 to 1300, "makes the Dust Bowl look like a picnic." This drought basically dried up all the rivers East of the Sierra Nevada Mountains. These are civilization-destroying, mega-droughts. One of those droughts "has been tied by some researchers to the decline of the Anasazi or Ancient Pueblo Peoples in the Colorado Plateau in the late 13th century."

The researchers' findings are unusually robust explained NASA's Cook: "The surprising thing to us was really how consistent the response was over these regions, nearly regardless

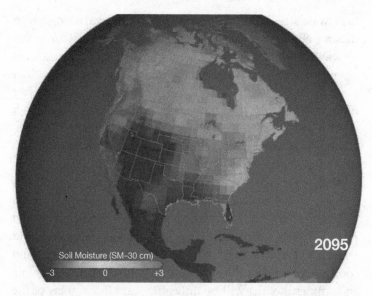

Figure 3.1 The National Aeronautics and Space Administration (NASA) study projections for all of North America if we stay our current, business-as-usual emissions path.
Source: NASA

of what model we used or what soil moisture metric we looked at. It all showed this really, really significant drying."

To summarize, several recent robust studies find that human-caused carbon emissions are putting large parts of the habited and arable land of the developed and developing world on track for the worst imaginable multidecadal droughts this century. Such Dust-Bowlification would be one of the most consequential impacts of climate change for the world.

What are the expected health impacts of climate change?

Climate change is expected to have a broad range of direct and indirect impacts on health this century. These impacts range from increased mortality due to longer and stronger heat waves to health problems created by warming-driven urban smog to risks posed by malnutrition and lack of access to water. Warming will have some beneficial impacts, most notably a modest drop in cold-related illness and death. "But globally over the 21st century, the magnitude and severity of negative impacts are projected to increasingly outweigh positive impacts," as the Intergovernmental Panel on Climate Change concluded in its comprehensive 2014 literature review on "Impacts, adaptation, and vulnerability."[31]

Most worrisome, if humanity stays near its current path of greenhouse gas emissions, the IPCC warns with "high confidence" that "the combination of high temperature and humidity in some areas for parts of the year is projected to compromise normal human activities, including growing food or working outdoors." In that case, simply being outdoors in summer months will be unhealthy, and those areas of the world would increasingly be seen as uninhabitable.

Although one might think that the human health impacts of global warming would be among the most well studied areas of climate change, it is only in the last decade that the medical community and other health professionals have focused on this issue in depth. As recently as 2009, a landmark Health Commission created by *The Lancet* medical journal and the University College London (UCL) Institute for Global Health could warn that the "full impact" of climate change to human health "is not being grasped by the healthcare community or policymakers." Lead author, Anthony Costello, a pediatrician and director of UCL Institute for Global Health, said that he himself "had not realised the full ramifications of climate change on health until 18 months ago."

The report, "Managing the Health Effects of Climate Change," concluded, "Climate change is the biggest global health threat of the 21st century." It warned, "Climate change will have devastating consequences for human health from":

- changing patterns of infections and insect-borne diseases, and increased deaths due to heat waves

- reduced water and food security, leading to malnutrition and diarrhoeal disease
- an increase in the frequency and magnitude of extreme climate events (hurricanes, cyclones, storm surges) causing flooding and direct injury
- increasing vulnerability for those living in urban slums and where shelter and human settlements are poor
- large-scale population migration and the likelihood of civil unrest

A 2011 editorial in The *British Medical Journal*, led by the surgeon rear admiral of the UK's Ministry of Defence, reviewed and synthesized recent reports on "Climate change, ill health, and conflict." The editorial warned that "Climate change poses an immediate and grave threat, driving ill health and increasing the risk of conflict, such that each feeds on the other." The threat posed by climate change to regional security "will limit access to food, safe water, power, sanitation, and health services and drive mass migration and competition for remaining resources." There will be a rise in starvation, diarrhea, and infectious diseases as well as in the death rate of children and adults. The authors note that "in 2004, seven of the 10 countries with the highest mortality rates in children under 5 were conflict or immediate post-conflict societies."

The warmer temperatures are, the more ozone smog that forms. The combination of increased carbon dioxide in the atmosphere and a warming climate will triple the percentage of the world population affected by low-level ozone, according to the UK's Met Office Hadley Center. They concluded that on our current emissions path, "By the 2090s close to one-fifth of the world's population will be exposed to ozone levels well above the World Health Organization recommended safe-health level." That would be some 2 billion people. A 2014 study by the U.S. National Center for Atmospheric Research found that Americans will experience a 70% increase in unhealthy ozone smog by mid-century—unless there are very strong regulations put in place to sharply cut the emissions of smog-forming pollutants.

Climate change harms air quality and human health in others ways. "Climate change is projected to increase the frequency of wildfire in certain regions of the United States," as the Congressional-mandated 2014 U.S. National Climate Assessment (NCA) notes. "Wildfire smoke contains particulate matter, carbon monoxide, nitrogen oxides, and various volatile organic compounds (which are ozone precursors) and can significantly reduce air quality, both locally and in areas downwind of fires." In fact, climate change is projected to increase the frequency of wildfires in many regions of the world, because up to one third of the inhabited land mass of the planet sees simultaneous warming and drying of many regions around the world. The potential health implications are huge. A 2012 study in Environmental Health Perspectives, "Estimated Global Mortality Attributable to Smoke from Landscape Fires," concluded the "annual global mortality attributable to landscape fire smoke" is a remarkable 339,000 deaths a year.

Many tropical diseases are tropical because their insect or animal host prefer warmer climates. A 2015 report on neglected tropical diseases by the World Health Organization found, "Climate variability and long-term climate changes in temperature, rainfall and relative humidity are expected to increase the distribution and incidence of at least a subset of these diseases." For instance, the World Health Organization notes, "Dengue has already re-emerged in countries in which it had been absent for the greater part of the last century." The 2014 NCA concurs, "Large-scale changes in the environment due to climate change and extreme weather events are increasing the risk of the emergence or reemergence of health threats that are currently uncommon in the United States, such as dengue fever."

The future is problematic, as parasite experts explain in an article, "Evolution in Action: Climate Change, Biodiversity

Dynamics and Emerging Infectious Disease" (EID). That article, which examines our "current EID crisis," is part of a special April 2015 theme issue on "Climate Change and Vector-Borne Diseases of Humans" in the UK Royal Society journal *Philosophical Transactions B*. "The appearance of infectious diseases in new places and new hosts, such as West Nile virus and Ebola, is a predictable result of climate change," as coauthor Daniel R. Brooks explains. Brooks says, "It's not that there's going to be one 'Andromeda Strain' that will wipe everybody out on the planet" (referring to a deadly fictional pathogen). However, he warns, "There are going to be a lot of localized outbreaks that put a lot of pressure on our medical and veterinary health systems. There won't be enough money to keep up with all of it. It will be the death of a thousand cuts."

Certainly there have been major advances in the fight against many tropical diseases, but those are primarily due to medical and public health advances. Climate change simply makes the job harder for all those focused on public health around the world.

How does global warming affect human productivity?

One of the most important but least discussed impacts of global warming is how it will affect human productivity, especially outdoors. In recent years, a number of studies have projected that global warming will have a serious negative impact on labor productivity this century, with a cost to society that may well exceed that of all other costs of climate change combined. One expert summed up the literature this way in 2011: "National output in several [non-agricultural] industries seemed to decline with temperature in a nonlinear way, declining more rapidly at very high daily temperatures."[32] Here is a look at what we know.

In 2013, a study from the NOAA projected that "heat-stress related labor capacity losses will double globally by 2050 with a warming climate." If we stay near our current greenhouse gas emissions pathway, then we face a potential 50% drop in labor capacity in peak (summer) months by century's end. Figure 3.2 is from that study, "Reductions in Labour Capacity from Heat Stress Under Climate Warming."

Staying on the business-as-usual emissions path (Representative Concentration Pathway [RCP]8.5) would risk sharp drops in labor productivity by 2100 and even deeper drops after that.

A 2010 paper for the U.S. National Bureau of Economic Research by Joshua Zivin and Matthew Neidell examined "Temperature and the Allocation of Time: Implications for Climate Change." That study determined "the number of

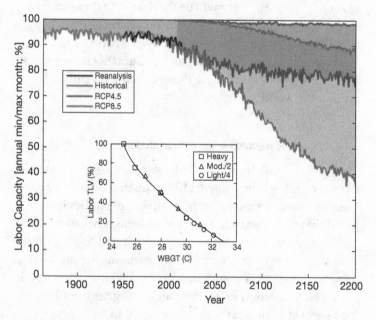

Figure 3.2 Individual labor capacity (%) during annual minimum (upper lines) and maximum (lower lines) heat stress months. RCP8.5 (red lines) is close to our current emissions path. These are derived from calculations of the wet bulb globe temperature (WBGT), which "is used as a measure of combined thermal and humidity stress on human physiology and activity."

CREDIT: NOAA

minutes in a day that individuals (who work in outdoor or temperature-exposed sectors in the USA) spent working as a function of maximum temperature (in Fahrenheit) that day." It found that productivity starts to nosedive at 90°F, and it collapses at 100°F. Andrew Gelman, director of the Applied Statistics Center at Columbia University, summed up the research this way in 2012: "2% per degree Celsius ... the magic number for how worker productivity responds to warm/hot temperatures." The negative impact seems to start at approximately 26°C (79°F).

This loss of productivity is by no means the most life threatening of climate impacts. People can, after all, simply stay indoors on the hottest and most humid days. However, it is one of the most important unmodeled climate impacts that makes the likely cost of climate change far higher than standard economic models suggest. If we stay near our current path of carbon pollution emissions, then, as we move toward mid-century, a larger and larger fraction of our summertime will be intolerable outside.

Stanford University made this point in a 2011 news release on research forecasting "permanently hotter summers":

> The tropics and much of the Northern Hemisphere are likely to experience an irreversible rise in summer temperatures within the next 20 to 60 years if atmospheric greenhouse gas concentrations continue to increase, according to a new climate study by Stanford University scientists ...
>
> "According to our projections, large areas of the globe are likely to warm up so quickly that, by the middle of this century, even the coolest summers will be hotter than the hottest summers of the past 50 years," said the study's lead author, Noah Diffenbaugh.

Here are the projected days above 100°F on our current emissions path (Figure 3.3), via the NCA.

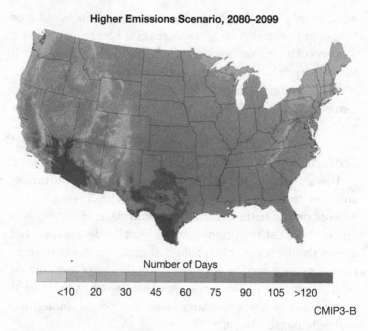

Higher Emissions Scenario, 2080–2099

Number of Days

<10   20   30   45   60   75   90   105   >120

CMIP3-B

Figure 3.3 Days above 100°F in higher emissions (business as usual) scenario.
Via the U.S. National Climate Assessment (NCA)

Absent deep reductions in global emissions, much of Kansas by century's end could well be above 100°F for nearly the whole summer. By century's end, much of the southern United States will see temperatures above 90°F for 5 months of the year or more, a big change from just the recent past, as Figure 3.4 from the U.S. National Climate Assessment shows.

As NOAA explained, outdoors, such warming eventually "eliminates all labor capacity in the hottest months in many areas, including the lower Mississippi Valley."

What does this mean for productivity? Professor Solomon M. Hsiang wrote in 2012:

In my 2010 PNAS paper, I found that labor-intensive sectors of national economies decreased output by roughly 2.4% per degree C and argued that this looked suspiciously like it came from reductions in worker output. Using a totally different method and dataset, Matt Neidell and Josh Graff Zivin found that labor supply in micro data fell by 1.8% per degree C. Both responses kicked in at around 26C.

Hsiang's own work shows "national output in several [non-agricultural] industries . . . declining more rapidly at very high daily temperatures." A Japanese professor found that indoors, "every degree rise in temperature above 25 Celsius (77 degrees Fahrenheit) resulted in a 2% drop in productivity," as the *New York Times* reported in 2012.

Thus, very different types of research using different data sets yield similar results. This research is essentially about adaptation—one key way that healthy people respond to high temperatures is simply to work less. NOAA notes an important caveat about its research, which tends to make the results conservative: "In focusing on the capacity of healthy, acclimated individuals, this study also severely underestimates heat stress implications for less-optimally acclimated individuals such as the young, old, and sick."

Hsiang points out the bottom line (emphasis added):

It's worth noting that reductions in worker output have never been included in economic models of future warming ... despite the fact that experiments fifty years ago showed that temperature has a strong impact on worker output.... In my dissertation I did some back-of-the-envelope estimates using the above numbers and found that productivity impacts alone might reduce per capita output by ~9% in 2080-2099 (in the absence of strong adaptation). **This cost exceeds the combined cost of all other projected economic losses combined.**

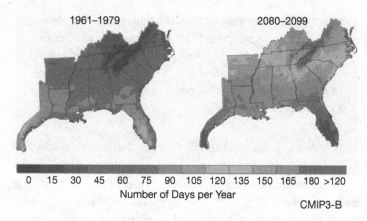

Figure 3.4 Via NCA.

The number of days per year with peak temperature over 90°F is expected to rise significantly, especially under a higher emissions scenario as shown in the map above. By the end of the century, projections indicate that North Florida will have more than 165 days (nearly 6 months) per year over 90°F, up from approximately 60 days in the 1960s and 1970s. The increase in very hot days will have consequences for human health, drought, and wildfires.

This suggests that standard projections of the economic cost from climate change may be low by more than a factor of two.

Does carbon dioxide at exposure levels expected this century have any direct impacts on human health or cognition?

It might seem the question of the direct impact of carbon dioxide on humans would be well studied and largely settled by now. Nonetheless, one of the single most important potential impacts of carbon pollution remains uncertain. Exceedingly few studies have measured the direct impact on human task performance of just raising carbon dioxide levels to 600 parts per million or 1000 ppm. However, some recent studies suggest that the level of carbon dioxide that humans are going to be routinely exposed to by century's end, especially indoors, could well diminish our decision-making performance.

Researchers at the Indoor Environment Group at Lawrence Berkeley National Laboratory (LBNL) and the State University of New York (SUNY) Upstate Medical University found that "relative to 600 ppm, at 1,000 ppm CO_2, moderate and statistically significant decrements occurred. in six of nine scales of decision-making performance."[33] That study built upon work by researchers in Hungary indicating that carbon dioxide levels of 3,000 ppm had observable negative impacts on some task performance. A subsequent LBNL study on ventilation rates, although not a replication of the original study, "is consistent with the results of [their] earlier work showing that the addition of pure carbon dioxide reduced decision-making performance." A 2015 study from Danish researchers examining impacts on task performance (rather than high cognitive function) found more mixed results, but concluded that "in the real world, high carbon dioxide levels, along with natural byproducts generated by the human body, may affect neurophysiology and task performance," as the lead author, Dr. Pawel Wargocki told me.

A fall 2015 study led by Dr. Joseph Allen at the Harvard School of Public Health confirmed the LBNL findings regarding CO_2 and decision-making performance, *but with effects seen at even lower concentrations of CO_2 (930 ppm)*. The Harvard study found that, on average, a typical participant's cognitive scores dropped 21% with a 400 ppm increase in CO_2 with 6 of 9 decision-making performance domains impacted at the 930 ppm level, and 8 of 9 negatively impacted at 1400 ppm.

A handful of studies cannot be considered definitive on an issue of such magnitude. However, as study coauthor Dr. William Fisk (leader of LBNL's Indoor Environment Group) explained to me, "anything that has the potential to reduce our cognitive performance even by a small amount deserves a great deal of study." That is especially true when we are talking about an impact to which billions of people will be exposed, potentially affecting future generations for many, many centuries. Given the preliminary research's implications for the future of humanity, scientists and medical professionals need to see whether these findings are replicable, if the effect persists with prolonged exposure to CO_2, and if any population subgroups might be especially vulnerable to higher levels of CO_2. We also must find out what is the lowest threshold level of CO_2 for negative impacts to be observed.

For most of human evolution and modern history, carbon dioxide levels in the atmosphere were in a fairly narrow range of 180 to 280 ppm. Also, during that time, most people spent most of their time outdoors or in enclosures that were not tightly sealed. We are now at 400 ppm and rising more than 2 ppm a year, a rate that is projected to rise faster and faster unless we sharply and quickly cut CO_2 emissions. We have been headed toward carbon dioxide levels in the atmosphere of 900 ppm or more. Recent commitments made by major countries leading up to the December 2015 Paris climate talks would take us off that pathway. On the other hand, as we have seen, there are many major carbon-cycle amplifying feedbacks that are not included in the climate models; therefore, even with a successful international climate agreement, we are still at risk for high CO_2 levels outdoors.

However, indoors is a different matter because humans generate and breathe out CO_2. This means that in buildings—the places where most people work and live—CO_2 concentrations are considerably higher than outdoors. In addition, that differential increases when more people are crammed into a space and when the ventilation is poor. As Dr. Allen told me, the higher outdoor CO_2 concentrations get, the higher building ventilation rates will need to be to keep indoor CO_2 concentrations low.

How high are CO_2 levels indoors? The LBNL article notes, "In surveys of elementary school classrooms in California and Texas, average CO_2 concentrations were above 1,000 ppm, a substantial proportion exceeded 2,000 ppm, and in 21% of Texas classrooms peak CO_2 concentration exceeded 3,000 ppm." In a sample of 100 offices, "5% of the measured peak indoor CO_2 concentrations exceeded 1,000 ppm," with an outside CO_2 level of some 400 ppm. One small study found that offices "where important decisions are sometimes made, can have elevated CO_2 concentrations—for example, up to 1,900 ppm during 30- to 90-min meetings."

The LBNL-led study also notes that we expect high CO_2 concentrations, "In some vehicles (aircraft, ships, submarines, cars, buses, and trucks), because of their airtight construction or high occupant density." In fact, the Harvard team has monitored CO_2 on commercial airplanes and found CO_2 concentrations are typically between 1300 and 1600 ppm, with a maximum recorded concentration of over 11,000 ppm! Some studies suggest that in cars where the ventilation system is closed or recirculating air, CO_2 levels can also be very high.

Clearly it is vital to find out whether people's decision-making ability or judgment is impaired by CO_2 levels in the range of 900 to 1000 ppm—or even lower. That is especially true because, as CO_2 levels for outside air rise this century, the CO_2 levels inside are also likely to rise. If the atmosphere does hit anywhere near, say, 700 ppm this century or next, then avoiding 900 to 1,000 ppm in a great many indoor settings would be extremely difficult. One 1997 study of *outdoor* urban settings observed CO_2 levels as much as 100 ppm above the ambient levels of the time.

Because humans are routinely exposed to 1000 ppm or more in situations in which judgment is critical, you might ask why we have not done extensive studies already to figure out whether that level does impair human decision-making. Dr. Fisk told me, for instance, we do know that low ventilation and high CO_2 levels in classrooms are associated with increased absenteeism and even poorer performance on standardized tests. The LBNL and SUNY study explains why the CO_2-only studies have not been done:

Prior research has found that with higher indoor levels of CO_2, indicating less outdoor air ventilation per person, people tend to be less satisfied with indoor air quality, report more acute health symptoms (e.g., headache, mucosal irritation), work slightly slower, and are more often absent from work or school. It has been widely believed that these associations exist only because the higher indoor CO_2 concentrations occur at lower outdoor air ventilation rates and are, therefore, correlated with higher levels of other indoor-generated pollutants that directly cause the adverse effects.

That is, people have assumed the cause of indoor air quality problems were due to volatile organic compounds (such as formaldehyde) and other pollutants known to cause harm that are generally correlated with high CO_2 levels. The result is that "CO_2 in the range of concentrations found in buildings (i.e., up to 5,000 ppm, but more typically in the range of 1000 ppm) has been assumed to have no direct effect on occupants' perceptions, health, or work performance."

What did the original LBNL study find when indoor CO_2 levels were raised and all other factors held constant? There were "statistically significant and meaningful reductions in decision-making performance" in the test subjects based on a standard assessment used for assessing cognitive function:

> At 1,000 ppm CO_2, compared with 600 ppm, performance was significantly diminished on six of nine metrics of decision-making performance. At 2,500 ppm CO_2, compared with 600 ppm, performance was significantly reduced in seven of nine metrics of performance, with percentile ranks for some performance metrics decreasing to levels associated with marginal or dysfunctional performance.

Dr. Wargocki told me the original LBNL study had an "exemplary design" with "systematic results" showing that "high cognitive skills were most affected" by high CO_2 levels. His team decided to repeat the experiment, but were unable to use tests that measured high cognitive skills. Instead they mainly looked at more basic task performance including typing, addition, and proofreading. Tests run for small groups over several hours at 1000 ppm and 3,000 ppm of CO_2, did not find statistically significant impact on basic task performance. There were "some indications performance was impacted at 3,000 ppm" on the most complicated psychological test. The Danish researchers also ran the elevated CO_2 tests a second time including the same kind of elevated levels of human bioeffluents you would expect to see in many indoor environments. They found evidence that performance was impacted at the 3,000 ppm level. They also measured the subjects' metabolic rate and respiration and saliva bio-markers (like cortisol). In the 3,000 ppm case with bioeffluents, researchers found indications of an increased stress response.

Unlike the Dutch study, the Harvard School of Public Health did use the same cognitive assessment tool as the LBNL study. The Harvard team used a robust 'double-blinded' study where neither participants nor the data analysts were aware of conditions in the simulated work environment on each day. Harvard researchers were interested in the impacts of longer exposures to CO_2 (a full work day) for adult office workers, and combined effects of CO_2, ventilation and volatile organic compounds. Despite any study design differences, the Harvard researchers report that "effects were consistent [with LBNL] a) in both study populations, [showing] that knowledge workers and students are equally impacted by CO_2 and b) at different exposure durations, indicating that even short elevated exposures can have effects on cognitive function."

The LBNL study, the follow up, the Danish study, the Harvard study, and the original Hungarian study are not definitive answers to the important question of the impact of higher carbon dioxide levels on human decision-making. We do not know, for instance, the impact of very prolonged exposure to higher CO_2 levels—although the Harvard study that used full day exposures points out, "The longer exposures in our study demonstrate that there is not a desensitization or compensatory response from longer exposures." We do not know exactly what role the other human bioeffluents might play when combined synergistically with elevated CO_2 levels. We do not know whether subgroups (i.e., children, the elderly, people with certain medical conditions) would be affected more than others. And we do not know what is the lowest threshold level of CO_2 for negative impacts to be observed. However, the potential impact of even a small effect on cognition and productivity would be so huge that, as the LBNL authors conclude, "Confirmation of these findings is needed."

What is ocean acidification and why does it matter to sea life?

One quarter of the carbon dioxide humans emit into the air gets absorbed in the oceans. The carbon dioxide that dissolves in seawater forms carbonic acid, which in turn acidifies the ocean. As a result, the oceans have acidified some 30% since the dawn of the industrial revolution, as measured by its dropping pH, a common measure of acidity.

The oceans are now acidifying faster than they ever have over the last 300 million years, during which time there were four major extinctions driven by natural bursts of carbon. A 2010 *Nature Geoscience* study, "Past Constraints on the Vulnerability of Marine Calcifiers to Massive Carbon Dioxide Release," explained that the oceans are now acidifying 10 times faster today than 55 million years ago when a mass extinction of marine species occurred. An April 2015 study in the journal *Science*, "Ocean Acidification and the Permo-Triassic Mass Extinction," found that the cause of an earlier mass extinction was rapidly acidifying oceans driven by a major pulse of carbon dioxide emissions into the atmosphere.

Why does ocean acidification threaten marine life? As carbon dioxide is absorbed in water, it causes chemical reactions that reduce "saturation states of biologically important calcium carbonate minerals," which "are the building blocks for the skeletons and shells of many marine organisms" (as NOAA explains). In the parts of the ocean teeming with life, the seawater has an overabundance (supersaturation) of these calcium carbonate minerals used by so-called "calcifying organisms," which include corals, mollusks, and some plankton. As the ocean absorbs more carbon dioxide, more and more places in the ocean are becoming undersaturated with these mineral, thereby threatening calcifying organisms. Besides a decline in calcification, the World Meteorological Organization explained in 2014, "Other impacts of acidification on marine biota include reduced survival, development and growth rates, as well as changes in physiological functions and reduced biodiversity."[34]

The 2015 *Science* study concluded that the Permo-Triassic extinction 252 million years ago, which is considered the "the greatest extinction of all time," happened during the time when massive amounts carbon dioxide were injected into the atmosphere, first slowly and then quickly (driven by volcanic eruptions). The researchers found that "During the second extinction pulse, however, a rapid and large injection of carbon caused an abrupt acidification event that drove the preferential loss of heavily calcified marine biota." How bad was this extinction? Besides killing over 90% of marine life, it wiped out some 70% of land-based animal and plant life.

Ocean acidification has long been a great concern of the world's climate scientists, in part because of its implications for global food production. In June 2009, some 70 Academies of Science issued a joint statement on ocean acidification. These groups of leading scientists from the major developed and developing countries warned "Ocean acidification is irreversible on timescales of at least tens of thousands of years" and "Marine food supplies are likely to be reduced with significant implications for food production and security in regions dependent on fish protein, and human health and wellbeing."

Today, coral reefs alone are estimated to support a quarter of all marine life. NOAA explains that "The fish that grow and live on coral reefs are a significant food source for half a billion people worldwide." The combination of warming waters and acidification have already caused serious harm to major coral reef around the world, and many appear unlikely to survive the century. Oceanographer and coral expert J.E.N. Veron (former chief scientist of the Australian Institute of Marine Science) has written, "The science is clear: Unless we change the way we live, the Earth's coral reefs will be utterly destroyed within our children's lifetimes."

Ocean acidification and carbon pollution have already proven to be a major threat to the U.S. oyster industry, as was clear from the "The Great Oyster Crash" of 2007 in coastal Oregon and Washington. There were "near total failures of developing oysters in both aquaculture facilities and natural ecosystems on the West Coast," as NOAA put it, with oyster larvae dying by the millions. Why? Originally it was thought that rapidly acidifying coastal waters made it difficult for larvae to build the shells needed for survival. However, a December 2014 *Nature Climate Change* study of Pacific oyster and Mediterranean mussel larvae determined that "the earliest larval stages are directly sensitive to saturation state, not carbon dioxide (CO_2) or pH" (acidity). So what matters most is how much calcium carbonate is in the ocean water relative to the total amount it could hold.

This finding has dramatic consequences for the speed at which rising carbon dioxide levels will harm ocean life. Lead author George Waldbusser, an Oregon State University marine ecologist and biogeochemist, explains why:

Larval oysters and mussels are so sensitive to the saturation state (which is lowered by increasing CO_2) that the threshold for danger will be crossed "decades to centuries" ahead of when CO_2 increases (and pH decreases) alone would pose a threat to these bivalve larvae. "At the current rate of change, there is not much more room for the waters off the Oregon coast to absorb more CO_2

without crossing the threshold we have identified with respect to saturation state," he said.

This finding suggests some of the worst impacts of rising carbon dioxide levels in the ocean may come sooner than expected.

What is biodiversity and how will climate change impact it?

Biodiversity is short for "biological diversity." It measures the variety of life on planet Earth. Some ecosystems are rich in biodiversity with a wide variety of flora and fauna, such as the Amazon rainforest or major coral reefs. A 2010 theme issue of the UK Royal Society journal *Philosophical Transactions B* on "Biological diversity in a changing world" concluded that "There are very strong indications that the current rate of species extinctions far exceeds anything in the fossil record." A major 2014 review article in the journal *Science* (led by Duke conservation ecologist Stuart Pimm), "The Biodiversity of Species and Their Rates of Extinction, Distribution, and Protection," concluded that "Current rates of extinction are about 1000 times the background rate of extinction. These are higher than previously estimated and likely still underestimated."[35]

The current mass extinction is due to a combination of factors, many driven by humans, including habitat destruction and overfishing and overhunting. Several aspects of climate change have begun contributing to species extinction, but what is of most concern to biologists today is that as the rate of global warming speeds up in the coming decades, the climate may well change too quickly for many if not most species to adapt.

We have already seen that the rate of ocean acidification is considerably faster today than during previous mass ocean extinctions, and that carbon pollution does far more to oceans than just acidify them. The Royal Society theme issue points out, "Tropical forests are repositories of a large fraction of the Earth's biological diversity. They are also being degraded at unprecedented rates." We have seen that the Amazon has recently experienced multiple 100-year droughts. The Amazon's dry season now lasts 3 weeks longer than it did 30 years ago. Multiple studies project that climate change will turn the normal climate of the Amazon into moderate to severe drought.

Determining how climate change will affect biodiversity is complicated by a number of factors. For instance, as Dr. Pimm said in 2014, "Most species remain unknown to science, and they likely face greater threats than the ones we do know." In addition, extinction is the result of a synergy of factors, many of which are directly attributable to humans (and therefore can be influenced by our behaviors). Of course, species can move and adapt, if the rate of change of the climate is not too fast. Moreover, humans are increasingly working to help species survive, even helping with "species migration and dispersal," as the Intergovernmental Panel on Climate Change changed noted in its 2014 Fifth Assessment. Even so, the IPCC warned that we are risking "substantial species extinction . . . with risk increasing with both magnitude and rate of climate change."

There is more to biodiversity than just the number of species, as shown in a 2011 article in *Nature Climate Change*, "Cryptic Biodiversity Loss Linked to Global Climate Change." This was the first global study "to quantify the loss of biological diversity on the basis of genetic diversity." Cryptic biodiversity "encompasses the diversity of genetic variations and deviations within described species." It could only be studied in detail because molecular-genetic methods were developed. Scientists with the German Biodiversity and Climate Research Centre noted that "If global warming continues as expected, it is estimated that almost a third of all flora and fauna species worldwide could become extinct." However, their research "discovered that the proportion of actual biodiversity loss should quite clearly be revised upwards: by 2080, more than 80% of genetic diversity within species may disappear in certain groups of organisms." Species may survive, but "the majority of the genetic variations, which in each case exist only in certain places, will not survive," as coauthor Dr. Carsten Nowak explained. A species' genetic variation increases its adaptability to a changing climate and changing habitats. Losing genetic diversity decreases the species' long-term chances for survival.

Finally, on the one hand, new technologies and strategies are making it easier for humans to protect endangered species. On the other hand, the rate of warming we face is so high that it will make it much harder for humans to protect endangered species, especially if the warming is so high that humans have to focus on feeding and protecting themselves.

How will climate change affect the agricultural sector and our ability to feed the world's growing population?

Feeding 9 billion or more people mid-century and beyond in the face of a rapidly worsening climate is likely to prove the greatest challenge the human race has ever faced. We are looking at the perfect storm of impacts. Dust-bowl conditions are projected to become the norm for large areas in both food-importing and food-exporting countries. This will be happening during a time when we will have drained many of the key aquifers that sustain agriculture in countries as diverse as India, China, and the United States. In addition, the glaciers that act as reservoirs for major river systems will be shrinking and vanishing, further reducing water availability during the crucial summer season for crops.

Every part of the world will be routinely hit by extreme deluges, floods, droughts, and heat waves that damage crops. At the same time, salt water intrusion from sea level rise threatens some of the richest agricultural deltas in the world, such as those of the Nile and the Ganges. Meanwhile, ocean acidification combined with ocean warming and overfishing may severely deplete the food available from the sea.

On the demand side, the United Nations Food and Agricultural Organization estimates that some 800 million people are chronically undernourished. In the coming decades, we will be adding another billion mouths to feed, then another billion and by most projections, another billion, taking us to 10 billion. At the same time, many hundreds of millions of people around the world will be entering the middle class, and, if they are anything like their predecessors around the globe, they will be switching from a mostly grain-based diet to a more meat-based one, which can require 10 times as much acreage and water per calorie delivered.

The World Bank issued an unprecedented warning about the threat to global food supplies in a 2012 report, "Turn Down the Heat: Why a 4°C Warmer World *Must* be Avoided." The Bank noted that the latest science was "much less optimistic" than what had been reported in the Intergovernmental Panel on Climate Change's 2007 Fourth Assessment report:

> **These results suggest instead a rapidly rising risk of crop yield reductions as the world warms.** Large negative effects have been observed at high and extreme temperatures in several regions including India, Africa, the United States, and Australia. For example, significant nonlinear effects have been observed in the United States for local daily temperatures increasing to 29°C for corn and 30°C for soybeans. **These new results and observations indicate a significant risk of high-temperature thresholds being crossed that could substantially undermine food security globally in a 4°C world.**

And that's just temperature rise: "Compounding these risks is the adverse effect of projected sea-level rise on agriculture in important low-lying delta areas." Moreover, we have the threat to seafood of ocean acidification. Finally, we have Dust-Bowlification:

> The report also says drought-affected areas would increase from 15.4% of global cropland today, to around 44% by 2100. The most severely affected regions in the next 30 to 90 years will likely be in southern Africa, the United States, southern Europe and Southeast Asia, says the report. In Africa, the report predicts 35% of cropland will become unsuitable for cultivation in a 5°C world.

What is some of the underlying science behind these conclusions? Using a "middle of the road" greenhouse gas emissions scenario, a study in *Science* found that for the more than five billion people who will be living in the tropics and subtropics by 2100, growing-season temperatures "will exceed the most extreme seasonal temperatures recorded from 1900 to 2006." The authors of "Historical Warnings of Future Food Insecurity with Unprecedented Seasonal Heat" conclude that "Half of world's population could face climate-driven food crisis by 2100."

A study led by MIT economists found that "the median poor country's income will be about 50% lower than it would be had there been no climate change." That finding was based on a 3°C warming by 2100, which is much less than the warming we are currently on track to reach. A further study led by NOAA scientists found that several regions would see rainfall reductions "comparable to those of the Dust Bowl era." Worse, unlike the Dust Bowl, which lasted about decade at its worst, this climate change would be "largely irreversible for 1,000 years after emissions stop." In other words, some of the most arable land in the world would simply turn to desert.[36]

In my *Nature* article, "The Next Dust Bowl," I wrote, "Human adaptation to prolonged, extreme drought is difficult or impossible. Historically, the primary adaptation to dust-Bowlification has been abandonment; the very word 'desert' comes from the Latin *desertum* for 'an abandoned place'." During the relatively short-lived U.S. Dust Bowl era, some 2.5 million people moved out of the Great Plains.

However, now we are looking at multiple, long-lived droughts and steadily growing areas of essentially nonarable land in the heart of densely populated countries and global breadbaskets. In a 2014 study, "Global warming and 21st century drying," the authors concluded, "An increase in evaporative drying means that . . . important wheat, corn and rice belts in the western United States and southeastern China, will be at risk of drought."

The study's lead author, Dr. Benjamin Cook, a top drought expert with joint appointments at NASA and Columbia, explained to me that we are headed into a "fundamental shift in Western hydro-climate." This drying includes the agriculturally rich Central Plains. The study warns that droughts in the region post-2050 "could be drier and longer than drought conditions seen in those regions in the last 1,000 years." Given how rapidly growing the population of the West is, I asked him whether there would be enough water for everyone there. He said "we can do it," but only "if you take agriculture out of the equation." However, that, of course, is not an option. Columbia University's Lamont-Doherty Earth Observatory further notes that "while bad weather periodically lowers crop yields in some places, other regions are typically able to compensate to avert food shortages. In the warmer weather of the future, however, crops in multiple regions could wither simultaneously." That would make food-price shocks "far more common," according to climatologist and study coauthor Richard Seager.

The international aid and development organization Oxfam has projected that global warming and extreme weather will combine to create devastating food price shocks in the coming decades. They concluded that wheat prices could increase by 200% by 2030 and corn prices could increase a remarkable 500% by 2030.

In 2014, the IPCC warned that humanity is risking a "breakdown of food systems linked to warming, drought, flooding, and precipitation variability and extremes." This was a key conclusion from its summary of what the scientific literature says about "Impacts, Adaptation, and Vulnerability," which every member government approved line by line. The IPCC pointed out that in recent years, "several periods of rapid food and cereal price increases following climate extremes in key producing regions indicate a sensitivity of current markets to climate extremes among other factors." So warming-driven drought and extreme weather have *already* begun to reduce food security.   ·

If we jump to a more heavily populated and climate-ravaged future, the IPCC warns that climate change will "prolong existing, and create new, poverty traps, the latter particularly in urban areas and emerging hotspots of hunger." You might think the question of the future of agriculture under high levels of warming would be something that has been well studied because of the importance of feeding so many people in a globally warmed world. However, the IPCC notes that "Relatively few studies have considered impacts on cropping systems for scenarios where global mean temperatures increase by 4°C [7°F] or more."

Even though humanity is currently headed towards 4°C [7°F] and beyond, we do not have a very good scientific picture of the full impact such climate change will have on agriculture and food supplies. The IPCC does mention briefly that our current path of unrestricted carbon emissions (the RCP8.5 scenario) holds unique risks for food supplies: "By 2100 for the high-emission scenario RCP8.5, the combination of high temperature and humidity in some areas for parts of the year is projected to compromise normal human activities, including growing food or working outdoors." If we warm anywhere near that much—some 4°C [7°F] or more—the challenge of feeding 9 billion people or more will become exponentially harder.

How is climate change a threat to national, regional, and global security?

Climate change will "increase risks of violent conflicts in the form of civil war and inter-group violence." That was a key summary conclusion of what the scientific literature says about climate "Impacts, Adaptation, and Vulnerability," as the U.N. Intergovernmental Panel on Climate Change reported in 2014. A landmark study from 2015 says climate change has done just that in Syria. And a 2014 U.S. Department of Defense study concluded, "Climate change ... poses immediate risks to U.S. national security," has impacts that can "intensify the challenges of global instability, hunger, poverty, and conflict" and will probably lead to "food and water shortages, pandemic disease, disputes over refugees and resources."[37]

The IPCC documents a large and growing literature on the connection between conflict and the kind of climate change we are facing. One of its central points is that

Climate change can indirectly increase risks of violent conflicts in the form of civil war and inter-group violence by amplifying well-documented drivers of these conflicts such as poverty and economic shocks. Multiple lines of evidence relate climate variability to these forms of conflict.

The link between climate change and violent conflict goes in both directions: "Violent conflict increases vulnerability to climate change. Large-scale violent conflict harms assets that facilitate adaptation, including infrastructure, institutions, natural resources, social capital, and livelihood opportunities." The threat to national security can be amplified by this vicious cycle whereby climate change makes violent conflict more likely and then that violent conflict makes a country more vulnerable to climate change. In that sense, climate change seems poised to help create many more of the most dangerous places on Earth: failed states.

In fact, climate change already appears to be driving and interacting with violent conflict. A 2015 study found that human-caused climate change was a major trigger of Syria's brutal civil war. The war that helped drive the rise of the terrorist Islamic State of Iraq and Greater Syria (ISIS) was itself spawned in large part by what one expert called perhaps "the worst long-term drought and most severe set of crop failures since agricultural civilizations began in the Fertile Crescent," from 2006 to 2010. That drought destroyed the livelihood of 800,000 people according to the U.N. and sent vastly more into poverty. The poor and displaced fled to cities, "where poverty, government mismanagement and other factors created unrest that exploded in spring 2011," as the study's news release explained.

The study, "Climate Change in the Fertile Crescent and Implications of the Recent Syrian Drought," published in *Proceedings of the National Academy of Sciences*, found that global warming made Syria's 2006 to 2010 drought two to three times more likely. "While we're not saying the drought caused the war," lead author Dr. Colin Kelley explained, "We are saying that it certainly contributed to other factors—agricultural collapse and mass migration among them—that caused the uprising."

"It's a pretty convincing climate fingerprint," Retired Navy Rear Admiral David Titley has said. Titley, also a meteorologist, said, "you can draw a very credible climate connection to this disaster we call ISIS right now." In particular, the study finds that climate change is already drying the region out in two ways: "First, weakening wind patterns that bring rain-laden air from the Mediterranean reduced precipitation during the usual November-to-April wet season. In addition, higher temperatures increased moisture evaporation from soils during the usually hot summers."

Tragically, this study and others make clear that for large parts of the not-terribly-stable region around Syria—including Lebanon, Israel, Jordan, and parts of Turkey and Iraq—brutal multiyear droughts are poised to become the norm in the coming decades if we do not reverse carbon pollution trends quickly.

Climate models had long predicted that the countries surrounding the Mediterranean would start drying out. In general, climate science says dry areas will get dryer and wet areas wetter. In 2011, a major NOAA study concluded that "human-caused climate change [is now] a major factor in more frequent Mediterranean droughts."

"The magnitude and frequency of the drying that has occurred is too great to be explained by natural variability alone," explained Dr. Martin Hoerling of NOAA's Earth System Research Laboratory, the lead author of the 2011 study (see Figure 3.5).

As previously discussed, large parts of the most inhabited and arable parts of the planet—the Southwest, Central Plains, the Amazon, southern Europe, much of Africa, southeastern

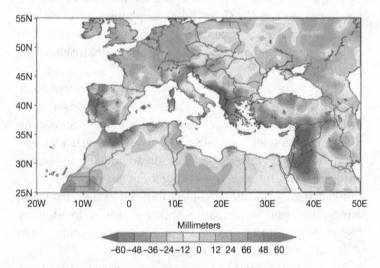

Millimeters

−60 −48 −36 −24 −12 0 12 24 66 48 60

Figure 3.5 Dark patches highlight lands around the Mediterranean that experienced significantly drier winters during 1971–2010 in comparison with the period of 1902–2010. Via NOAA.

China—face even worse heating and drying than what has already affected the Mediterranean.

So when might we expect warming-driven conflict to start to happen on a bigger scale? Because conflict has many contributing causes and typically requires some sort of political trigger, predicting exactly when we might see more conflict by climate change is difficult to do. In 2008, Thomas Fingar, then "the U.S. intelligence community's top analyst," estimates that it will happen by the mid-2020s, as "droughts, food shortages and scarcity of fresh water will plague large swaths of the globe, from northern China to the Horn of Africa." This "will trigger mass migrations and political upheaval in many parts of the developing world." The UK government's chief scientist, Professor John Beddington, laid out a scenario similar in a 2009 speech. He warned that by 2030, "A 'perfect storm' of food shortages, scarce water and insufficient energy resources threaten to unleash public unrest, cross-border conflicts and mass migration as people flee from the worst-affected regions," as the UK's Guardian put it.

What is the plausible best-case scenario for climate change this century?

The plausible best-case scenario for climate change this century would be keeping total warming below 2°C (3.6°F). That likely requires stabilizing atmospheric concentrations of carbon dioxide below 450 parts per million. Because we are already at 400 ppm and rising more than 2 ppm a year, and because concentrations of CO_2 in the air will not stop rising until we cut global emissions of CO_2 to 80% or more below current levels, that would require an aggressive worldwide effort.

The Intergovernmental Panel on Climate Change in its Fifth Assessment of the scientific literature developed some Representative Concentration Pathways (RCPs) to model future warming projections depending on how well we are able to control greenhouse gas emissions and concentrations. Their best-case scenario, RCP2.6, is one that provides a high likelihood of keeping total warming below 2°C. In that scenario, CO_2 concentrations peak around mid-century and then actually decline back to approximately 400 ppm by century's end. Getting atmospheric CO_2 levels to decline requires bringing net human-caused carbon pollution emissions below zero—that is, we are pulling more carbon dioxide out of the air than we are putting in. Because capturing atmospheric CO_2 and storing it somewhere permanently at large scale is not currently practical (see Chapter Six), whether this is a plausible scenario is open to interpretation. In principle, however, based on what we know today, nothing makes it impossible in the decades to come; therefore, as difficult and expensive as it might be to achieve, it can still be considered a possible best-case scenario, with a projected warming of 1.6°C (2.9°F).

By way of comparison, RCP8.5 approximates the business-as-usual pathway, where no significant measures are taken to cut global carbon pollution. In that case, CO_2 concentrations exceed 900 ppm by century's end. Warming above preindustrial levels hits about 4.2°C (8°F) by 2100, and temperatures keep rising because concentrations have not been stabilized. On the one hand, the emissions cuts already pledged in 2015 by China, the United States, the European Union, and other countries mean we are likely to veer off this path soon, if we have not already. On the other hand, the IPCC does not model key carbon-cycle feedbacks, such as the melting permafrost, so there remains a real risk we will still come close to RCP8.5 without further commitments by leading emitters.

Figure 3.6 is the graphic the IPCC published in its Fifth Assessment report comparing the two scenarios' projected future warming side-by-side.

In the IPCC best-case scenario, total warming from preindustrial levels would be kept to under 2°C over most of the heavily populated landmass of the world. In addition, compared with

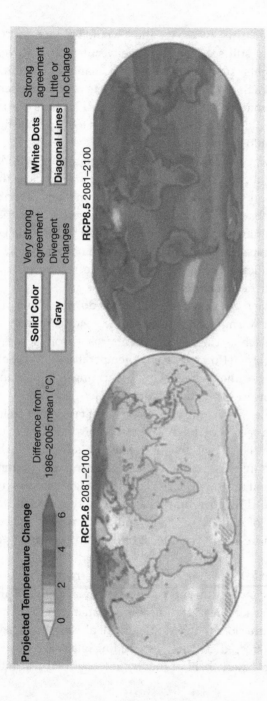

Figure 3.6 Humanity's choice via IPCC: Super-aggressive climate action starting immediately minimizes future warming and impacts (RCP2.6). Continued inaction (or the initiation of significant carbon-cycle feedbacks) results in very high levels of warming over many of the most populated areas of the world (RCP8.5).

current temperatures, future warming would be closer to 1°C. Yes, we would still see more extreme weather—heat waves, floods, droughts, superstorms—but the chances of turning the climate of a third of Earth's landmass into near-permanent Dust Bowl would be dramatically reduced. Impacts across the board—from acidification to species loss to human health impacts—would be sharply reduced.

This pathway would limit the worst economic impacts. As noted earlier, a 2013 NOAA study projected staying near the RCP8.5 pathway risks a potential 50% drop in labor capacity in peak months by century's end. The same study notes that "Only by limiting global warming to less than 3°C (5°F) do we retain labor capacity in all areas in even the hottest months." This is significant because productivity losses potentially exceed all other economic losses combined.

Another key point is that we really do not know the exact level of warming that prevents the major carbon cycle feedbacks—such as the melting permafrost—from severely complicating any effort to stabilize temperature and global climate. We do know that as total warming hits 2°C and then rises towards 3°C, the permafrost defrosts at a faster and faster rate. That in turn would require faster and faster reductions in carbon pollution. We also know that the higher CO_2 levels in the air get, the more likely it is that all of the major carbon sinks (particularly the oceans and soil) become less effective at taking up CO_2. That is one key reason why the world's leading governments along with the overwhelming majority of climate scientists and academies of science worldwide have set 2°C as a level of warming we must not exceed.

Perhaps the biggest uncertainty about this pathway is whether it is enough to stop collapse of a major part of the West Antarctic Ice Sheet. The latest observations and analysis find that we are close to that tipping point already, if we have not already crossed it. However, leading glaciologists and sea-level rise experts believe that by keeping total warming as low as possible, we slow the rate of sea rise and increase chances that we do not cross other key tipping points, including the loss of the Greenland Ice Sheet and a big part of the East Antarctic Ice Sheet. Certainly a rise of, say, 2 to 3 feet by 2100 rising a few inches a decade after that would be a dangerous climate impact, but it would be far less catastrophic than a 4 to 6 feet (or more) rise followed by an additional foot of sea rise every decade after that.

What is the plausible worst-case scenario for climate change this century?

The overwhelming majority of scientific research on climate change is not about the worst-case scenario. The Intergovernmental Panel on Climate Change, in its more than one-quarter century of existence, has never plainly laid out what that worst-case scenario is and what it would mean for human society. However, its most recent review of the scientific literature gave an upper range for the business-as-usual warming we would see by 2100 of a catastrophic 7.8°C (14°F)—if the climate response is at the high end of the estimated range, which is more likely than it being at the low end range. Moreover, none of the world's major scientific bodies have laid out the worst-case scenario either, although the UK's Royal Society has come closer than most.

In most aspects of our lives, however, humans—at an individual and societal level—are very risk averse, particularly when it comes to life-changing or catastrophic or irreversible risks. That is why we buy fire insurance even though the chances of losing your home to fire are quite small (unless you live in a known wildfire zone). That is also why we buy catastrophic health insurance—not because we expect to get cancer, but because we know that if we do and do not have insurance, the result might be bankruptcy on top of the illness. Worst-case scenario planning has driven a considerable amount of government spending for the military and epidemic-prevention. What little economic analysis has been done in this area suggests that worst-case scenarios for climate change—a scenario with relatively low probability and high catastrophic damages—can potentially dominate calculations of things such as the "social cost of carbon" or the total cost to humanity of climate inaction.

The scenario described in this chapter would probably be described by most as "high catastrophic damages," but it is merely close to the business-as-usual set of impacts according to the most recent observations and scientific analysis. These include upwards of 4°C (7°F) global warming, widespread drought and Dust-Bowlification, mass species loss on land and sea, increase in the most extreme type of weather events globally (including heat waves and superstorms), sea-level rise much greater than 6 feet by century's end with seas rising up to a foot a decade after that, the resulting increases in salt water infiltration and storm surges globally, and all these effects combining synergistically together to provide a myriad of threats to human health, national security, and our ability to feed a population headed toward 10 billion people. The generally cautious and conservative IPCC warned that that these impacts would lead to a "breakdown of food systems," more violent conflicts, and ultimately threaten to make some currently habited and arable land virtually unlivable for parts of the year.

In addition, we also face (1) the destruction of much of the permafrost and Amazon carbon sink, (2) the potential for the continued clogging up and weakening of other key carbon sinks, such as the oceans and soils, and (3) longer term, the thawing of the methane hydrates, which are frozen methane crystals under the permafrost and in the ocean. However, even though we know they are at risk, many of these key carbon cycle feedbacks (such as the loss of the massive amounts of carbon from the permafrost) are not even found in the most widely used climate models.

One way to look at the worst-case scenario is that instead of these impacts happening around 2100 or shortly thereafter, they could occur decades sooner if unmodeled carbon-cycle feedbacks kick in. In 2010, the Royal Society devoted a special issue of *Philosophical Transactions A* to look at this 4°C (7°F) scenario, "Four Degrees and Beyond: The Potential for a Global Temperature Increase of Four Degrees and Its Implications." That issue notes: "In such a 4°C world, the limits for human adaptation are likely to be exceeded in many parts of the world, while the limits for adaptation for natural systems would largely be exceeded throughout the world." The loss of natural systems would in turn make life for humans far more problematic than climate modelers have previously thought. The UK's *Guardian* describe this picture of a 4°C world as "a hellish vision," explaining "A 4C rise in the planet's temperature would see severe droughts across the world and millions of migrants seeking refuge as their food supplies collapse."[38]

The IPCC's 2014 Synthesis Report ties all the scientific literature together:

> In most scenarios without additional mitigation efforts ... warming is more likely than not to exceed 4°C [7°F] above pre-industrial levels by 2100. The risks associated with temperatures at or above 4°C include substantial species extinction, global and regional food insecurity, consequential constraints on common human activities, and limited potential for adaptation in some cases (*high confidence*).

Again, this 4°C world is not the plausible worst-case, it is the expected outcome of the emissions pathway we are currently on. The worst-case version would be if the 4C world occurred far sooner than expected. Dr. Richard Betts, Head of Climate Impacts at the Met Office Hadley Centre, laid out the "plausible worst case scenario" as lead author of one of the Royal

Society articles, "When Could Global Warming Reach 4°C?"
It contains this remarkable finding. If we stay near the high
emissions pathway, and "If carbon-cycle feedbacks are stronger [than currently modeled], which appears less likely but
still credible, then 4°C warming could be reached by the early
2060s in projections that are consistent with the IPCC's 'likely
range'."

The IPCC has very little to say about the catastrophic
impacts to the food system in the business-as-usual case
where the Earth warms 4°C to 5°C (7°F to 9°F), and it has nothing to say about even higher warming. It explains, "Relatively
few studies have considered impacts on cropping systems for
scenarios where global mean temperatures increase by 4°C
[7°F] or more." However, if we stay anywhere near our current
emissions path, or if carbon-cycle feedbacks are stronger than
currently modeled, we will continue to warm up past 4°C and
hit 6°C warming or more.

There has not been much modeling at all of what temperatures that high would mean for Homo sapiens. NOAA did
explore the impact of that kind of heat stress on productivity
in a 2013 study:

> Global warming of more than 6°C (11°F) eliminates all
> labor capacity in the hottest months in many areas,
> including the lower Mississippi Valley, and exposes most
> of the US east of the Rockies to heat stress beyond any-
> thing experienced in the world today. In this scenario,
> heat stress in NYC exceeds present day Bahrain, and
> Bahrain heat stress would induce hyperthermia in even
> sleeping humans.

By century's end, RCP8.5 would likely mean global warming
exceeding 4°C. The IPCC warns of "consequential constraints
on common human activities," explaining that "by 2100 for
RCP8.5, the combination of high temperature and humidity in

some areas for parts of the year is expected to compromise common human activities, including growing food and working outdoors." As we go past 4°C warming, we put ourselves at greater and greater risk of making large parts of the planet's currently arable and populated land (1) virtually uninhabitable for much of the year and (2) irreversibly so for hundreds of years.

Whether most species could survive in that scenario of rapid warming is problematic. In its coverage of the April 2015 study "Ocean Acidification and the Permo-Triassic Mass Extinction," led by Matthew Clarkson, the journal *Nature* reported, "The Great Dying might represent a worst-case scenario for the future if CO_2 emissions continue to rise, says Clarkson." That extinction killed over 90% of marine life (as well as 70% of land-based animal and plant life).

This section has focused on the warming this century, but the ultimate level of warming we are risking is considerably higher. In 2011, *Science* published a major review and analysis of paleoclimate data, "Lessons from Earth's Past" by National Center for Atmospheric Research scientist Jeffrey Kiehl. The study notes that "continuing on a business-as-usual path of energy use based on fossil fuels will raise [carbon dioxide levels in the air] to 900 to 1100 ppmv by the end of this century." It examines temperature reconstructions from the last time carbon dioxide hit 1000 ppmv, some 35 million years ago. The paper concludes, "an increase of CO_2 from 300 ppmv to 1000 ppmv warmed the tropics by 5° to 10°C and the Polar Regions by even more (i.e., 15° to 20°C)." On average, the Earth was 29°F (16°C) hotter the last time carbon dioxide levels were where they are headed. Kiehl concludes the following:

Earth's CO_2 concentration is rapidly rising to a level not seen in 30 to 100 million years, and Earth's climate was extremely warm at these levels of CO_2. If the world reaches such concentrations of atmospheric CO_2, positive feedback processes can amplify global warming beyond current modeling estimates. The human species and global ecosystems will be placed in a climate state never before experienced in their evolutionary history and at an unprecedented rate. Note that these conclusions arise from observations from Earth's past and not specifically from climate models. Will we, as a species, listen to these messages from the past in order to avoid repeating history?

The impacts of such warming are difficult to imagine. If post-2100, we were to get anywhere near the kind of warming that Kiehl's analysis of the paleoclimate data suggests, we could render large tracts of the planet simply uninhabitable for much of the year. That was the conclusion of a 2010 paper coauthored by Matthew Huber, professor of Earth and Atmospheric Sciences at Purdue. Huber explains the study's worst-case bottom line. A "21-degree [F] warming would put half of the world's population in an uninhabitable environment." He concluded the following:

"When it comes to evaluating the risk of carbon emissions, such worst-case scenarios need to be taken into account. It's the difference between a game of roulette and playing Russian roulette with a pistol. Sometimes the stakes are too high, even if there is only a small chance of losing."

What do scientists mean by "irreversible impacts" and why are they such a concern with climate change?

Most environmental problems that people, communities, and governments have experience dealing with are reversible. A polluted lake or river can be cleaned up and then used for swimming and fishing. A city with polluted air can put in place clean air standards and turn its brown haze into blue skies.

However, climate change is different from most environmental problems. The scientific literature has made it increasingly clear that key impacts are irreversible on a time scale of centuries and possibly millennia. This means that climate change creates risks that are unparalleled in human history. It also means that if we follow the traditional way of dealing with an environmental problem, that is, wait until the consequences are obvious and unmistakable to everybody, it will be "too late" to undo those consequences for a long, long time. Climate inaction inherently raises issues of equity because it will harm billions of people who have contributed little or nothing to the problem. However, what makes the issue unique in the annals of history is that the large-scale harm is irreparable on any timescale that matters (and that we could avoid the worst of the irreparable harms at a surprisingly low net cost, as discussed in Chapter Four).

Because irreversibility is such a unique and consequential fact about climate change, the world's leading climate scientists (and governments) took extra measures to emphasize the issue in the most recent international assessment of climate science by the U.N. Intergovernmental Panel on Climate Change—the November 2014 full, final "synthesis" report in its Fifth Assessment all of the scientific and economic literature. In the IPCC's final "synthesis" report of its Fourth Assessment, issued in 2007, irreversibility was only mentioned two times and there was minimal discussion in the Summary for Policymakers. Seven years later, the "Summary for Policymakers" of the IPCC's synthesis report mentions "irreversible" 14 times and has extended discussions of exactly what it means and why it matters. The full report has an even more detailed discussion.

What do the world's leading scientists mean by "irreversible impacts"? In the latest IPCC report, they explain that

Warming will continue beyond 2100 under all RCP scenarios except RCP2.6 [where emissions are cut sharply].

Surface temperatures will remain approximately constant at elevated levels for many centuries after a complete cessation of net anthropogenic CO_2 emissions.
A large fraction of anthropogenic climate change
resulting from CO_2 emissions is irreversible on a
multi-century to millennial time scale, except in the
case of a large net removal of CO_2 from the atmosphere
over a sustained period. . . .

It is *virtually certain* that global mean sea-level rise
will continue for many centuries beyond 2100, with the
amount of rise dependent on future emissions.

In other words, impacts will be much worse than described
in this report *after* 2100 in every case but the one where we
sharply cut carbon dioxide starting now (to stabilize at below
2°C total warming). In addition, whatever temperature the
planet ultimately hits thanks to human-caused warming, that
is roughly as high as temperatures will stay for hundreds of
years *after* we bring total net human-caused carbon pollution
emissions to zero.

The "case of a large net removal of CO_2 from the atmosphere
over a sustained period" means a time far beyond when
humanity has merely eliminated total net human-caused
emissions—from deforestation and burning fossil fuels (and
from whatever amplifying carbon-cycle feedbacks we have
caused, such as defrosting permafrost). To start reversing the
irreversible, we have to go far below zero net emissions to actually sucking vast quantities of diffuse CO_2 out of the air and
putting it someplace that is also permanent, which, according
to a 2015 National Academy of Sciences report (discussed in
Chapter Six), we currently do not know how to do on a large
scale. One can envision such a day when we might be able to
go far below zero—if we sharply reduce net carbon pollution
to zero by 2100, as we must to stabilize near 2°C. However, it is
much more difficult to imagine when it would happen if emissions are anywhere near current levels by 2100, and we have

started one or more major amplifying carbon-cycle feedbacks that make the job of getting to even zero net emissions doubly difficult.

If we do not get on the 2°C path, then some of the most serious climate changes caused by global warming could last a thousand years or more. The IPCC explained in 2014, "Stabilisation of global average surface temperature does not imply stabilisation for all aspects of the climate system." That is to say, as we warm above 2°C, then even at a point many hundreds of years from now when temperatures start to drop, some changes in the climate—sea-level rise being the most obvious example—will likely keep going and going.

The IPCC reports are primarily reviews of the scientific literature, so the new focus on the irreversible nature of climate change is no surprise. In a 2009 study titled "Irreversible Climate Change Because of Carbon Dioxide Emissions," researchers led by NOAA scientists concluded that "the climate change that is taking place because of increases in carbon dioxide concentration is largely irreversible for 1,000 years after emissions stop." It is significant to note that the NOAA-led study warned that it was not just sea-level rise that would be irreversible:

> Among illustrative irreversible impacts that should be expected if atmospheric carbon dioxide concentrations increase from current levels near 385 parts per million by volume (ppmv) to a peak of 450-600 ppmv over the coming century are irreversible dry-season rainfall reductions in several regions comparable to those of the **"dust bowl"** era and inexorable sea level rise.

Recent studies strongly support that finding for both sea-level rise and Dust-Bowlification of some of the world's most productive agricultural lands, as we have seen.

This 2014 Synthesis report may be the first time the world's leading scientists and governments explain why the irreversibility of impacts makes inaction so uniquely problematic. Here is the key finding (emphasis in original):

Without additional mitigation efforts beyond those in place today, and even with adaptation, warming by the end of the 21st century will lead to high to very high risk of severe, widespread, and irreversible impacts globally (*high confidence*). Mitigation involves some level of co-benefits and of risks due to adverse side-effects, but these risks do not involve the same possibility of severe, widespread, and irreversible impacts as risks from climate change, increasing the benefits from near-term mitigation efforts.

Why is this conclusion so salient? The IPCC is acknowledging that mitigation efforts taken to reduce greenhouse gas emissions have risks in addition to their cobenefits—"possible adverse side effects of large-scale deployment of low-carbon technology options and economic costs," as the full report puts it. However, the risks involved in reducing emissions are both quantitatively and qualitatively different than the risks deriving from inaction because they are not likely to be anywhere near as "severe, widespread, and irreversible."

The full 2014 "Synthesis" report expands on this point, noting that "Climate change risks may persist for millennia and can involve very high risk of severe impacts and the presence of significant irreversibilities combined with limited adaptive capacity." In sharp contrast, "the stringency of climate policies can be adjusted much more quickly in response to observed consequences and costs and create lower risks of irreversible consequences." Put another way, if some aspect of the emissions reduction strategy turns out to start having unexpected, significant negative consequences, humanity can quickly adjust to minimize costs and risks. However, inaction—failing to embrace strong mitigation—will lead to expected climate impacts that are not merely very long lasting and irreversible, but potentially beyond adaptation. For instance, sea-level rise would become so great, so rapid, and so unstoppable that we simply have to abandon the vast majority of coastal cities.

4

AVOIDING THE WORST
IMPACTS

This chapter will examine the 2°C warming target. It will explain why the major governments and scientific associations have embraced it as the limit for minimizing/avoiding dangerous climate change. It will take a big-picture perspective on how to avoid the worst impacts—including discussions of adaptation and geoengineering.

What is the biggest source of confusion about what humanity needs to do to avoid the worst climate impacts?

Perhaps the biggest source of confusion in the public climate discussion is that avoiding catastrophic warming requires stabilizing carbon dioxide *concentrations* not *emissions*. Studies find that many, if not most, people are confused about this, including highly informed people, and they mistakenly believe that if we stop increasing emissions, then global warming will stop. In fact, very deep reductions in greenhouse gas (GHG) emissions are needed to stop global warming. One study published in *Climatic Change* on the beliefs of Massachusetts Institute of Technology (MIT) graduate students, found that "most subjects believe atmospheric GHG concentrations can be stabilized while emissions into the atmosphere continuously exceed the removal of GHGs from it." The author, Dr. John Sterman from MIT's Sloan School of Management, notes that these beliefs are "analogous to arguing a bathtub filled faster than it drains will never overflow" and "support wait-and-see policies but violate conservation of matter."

Let me expand on the bathtub analogy. Although atmospheric concentrations (the total stock of CO_2 already in the air) might be thought of as the water level in the bathtub, emissions (the yearly new flow into the air) are represented by the rate of water flowing into a bathtub from the faucet. There is also a bathtub drain, which is analogous to the so-called carbon "sinks" such as the oceans and the soils. The water level will not drop until the flow through the faucet is less than the flow through the drain.

Similarly, carbon dioxide levels will not stabilize until human-caused emissions are so low that the carbon sinks can essentially absorb them all. Under many scenarios, that requires more than an 80% drop in CO_2 emissions. If the goal is stabilization of temperature near or below the 2°C (3.6°F) threshold for dangerous climate change that scientists and governments have identified, then carbon dioxide emissions need to approach zero by 2100. A key related point of confusion is that temperatures do not stop rising once atmospheric carbon dioxide levels have stabilized. It takes a while for the Earth's climate system to actually reach its equilibrium temperature for a given level of CO_2. If CO_2 levels stopped rising now, temperatures would keep rising for another few decades, albeit slowly. Put another way, the warming that we have had to date is due to CO_2 levels from last century. As long as we keep putting enough carbon dioxide into the air to increase CO_2 levels, then this lag will persist and the ultimate warming we face will continue to rise. In addition, certain key impacts, such as the disintegration of the great ice sheets, will also not stop for decades. Moreover, if we wait too long and pass the point of no return, then ice sheet collapse and sea-level rise will continue for centuries, even if temperatures stop rising.[39]

The MIT study, "Understanding Public Complacency About Climate Change: Adults' Mental Models of Climate Change

Violate Conservation of Matter," notes that there is an apparent "contradiction" in "public attitudes about climate change":

> Surveys show most Americans believe climate change poses serious risks but also that reductions in greenhouse gas (GHG) emissions sufficient to stabilize atmospheric GHG concentrations or net radiative forcing can be deferred until there is greater evidence that climate change is harmful. US policymakers likewise argue it is prudent to wait and see whether climate change will cause substantial economic harm before undertaking policies to reduce emissions. Such wait-and-see policies erroneously presume climate change can be reversed quickly should harm become evident, underestimating substantial delays in the climate's response to anthropogenic forcing.

Such a misconception of climate dynamics may lead some people to mistakenly believe that action to reduce carbon dioxide emissions does not need to start imminently.

What is the United Nations Framework Convention on Climate Change?

The United Nations Framework Convention on Climate Change (UNFCCC) is a global treaty on the environment approved by all of world's leading countries. It has become the primary venue for international climate negotiations, with 196 member nations or "parties" to the convention as of 2014. The UNFCCC was agreed upon at the June 1992 Rio Earth Summit.

The goal of the treaty was to set up an international process to "stabilize greenhouse gas concentrations in the atmosphere at a level that would prevent dangerous anthropogenic [human-caused] interference with the climate system." The UNFCCC did not define what that level was at the time. At the 2009 Conference of the Parties (COP) held in Copenhagen,

Denmark, the nations of the world agreed that 2°C was the threshold at which dangerous interference began. The original Convention itself had no binding targets for GHGs and no enforcement mechanism, but it did set up a legally non-binding target that called for the developed countries to bring their emissions of GHGs back to 1990 levels. The UNFCCC is primarily a framework by which the members can negotiate treaties ("protocols") that could be binding. It has held a COP for such negotiations every year from 1995.

The signatories to the Convention acknowledged "that the largest share of historical and current global emissions of greenhouse gases has originated in developed countries, that per capita emissions in developing countries are still relatively low and that the share of global emissions originating in developing countries will grow to meet their social and development needs." The Convention recognized the "common but differentiated responsibilities and respective capabilities" of each nation and established a core principle: "Accordingly, the developed country parties should take the lead in combating climate change and the adverse effects thereof."

That is why in 1997, in the Kyoto, Japan COP, the industrialized countries negotiated the Kyoto Protocol, which set targets and timetables only for the emissions of rich countries. Almost every industrialized nation in the world ratified the Protocol, with the notable exception of the United States. The Protocol required developing countries to cut total emissions of major GHGs 5% (or more) compared with 1990 levels by the 2008–2012 timeframe. The Protocol did lead to actions by many countries to cut carbon pollution, most notably the European Union. However, the absence of the United States coupled with rapid growth in developing countries' emissions post-2000, particularly China's, meant that overall global emissions continue to grow. The Paris COP in December 2015 is where the Parties have committed to develop a follow-on treaty to Kyoto, one that includes commitments by the United

States as well as major emitters in the developing world such
as China.

Why did scientists and governments decide 2°C (3.6°F) was the limit beyond which climate change becomes "dangerous" to humanity?

In December 2009, many of the world's leading nations recognized "the scientific view that the increase in global temperature should be below 2 degrees Celsius." In this Copenhagen
Accord, they stated, "We agree that deep cuts in global emissions are required according to science, and as documented
by the IPCC Fourth Assessment Report with a view to reduce
global emissions so as to hold the increase in global temperature below 2 degrees Celsius." At the December 2010 COP in
Cancun, the U.N. Framework Convention on Climate Change
officially embraced the goal of holding total global warming
to below 2°C above preindustrial levels.[40]

The original idea for a limit of 2°C dates back almost four
decades, to a 1977 study by Yale economics professor William
Nordhaus, "Economic Growth and Climate: The Carbon
Dioxide Problem." That paper looked at future emissions in
the case of business as usual ("uncontrolled") carbon dioxide emissions and concluded, "It appears that the uncontrolled path will lead to very large increases in taking the
climate outside of any temperature pattern observed in the
last 100,000 years." The paper identified 2°C as the "estimated
maximum [climate pattern variation] experienced" during
that time. The paper noted that the uncontrolled path would
lead to more than 4°C warming by 2100. All of these conclusions have been supported by an increasing amount of scientific research in recent years.

The 2°C limit "was contested diplomatically for over
13 years and was subject to different levels of scientific and
political criticism prior to its adoption at Copenhagen in 2009,"
explain the scientists at Climate Analytics, a nonprofit climate

science and policy institute, in their 2014 history of the goal. It was a build-up of scientific evidence, as documented in the various Intergovernmental Panel on Climate Change reports, especially the Fourth Assessment Report (AR4) in 2007, which won the Nobel Peace Prize and ultimately created the political consensus for action. Stefan Rahmstorf, Co-Chair of Earth System Analysis at the Potsdam Institute for Climate Impact Research and a sea-level rise expert, elaborated in 2014:

> One of the rationales behind 2°C was the AR4 assessment that above 1.9°C global warming we start running the risk of triggering the irreversible loss of the Greenland Ice Sheet, eventually leading to a global sea-level rise of 7 meters. In the AR5, this risk is reassessed to start already at 1°C global warming. And sea-level projections of the AR5 are much higher than those of the AR4.

In addition, since the Fifth Assessment Report, concern has grown sharply that we are near, or at, a tipping point for the great ice sheets, as discussed in Chapter Three. In May 2014, we learned that the West Antarctic Ice Sheet (WAIS) appears close to, if not past, the point of irreversible collapse, and we learned that "Greenland's icy reaches are far more vulnerable to warm ocean waters from climate change than had been thought." Late in 2014, observations revealed that Greenland and WAIS more than doubled their rate of ice loss in the previous 5 years. Then in 2015, researchers reported that a large glacier in the East Antarctic Ice Sheet turns out to be as unstable and as vulnerable to melting from underneath as WAIS is.

In his October 2014 analysis of the limit, Rahmstorf notes, "If anything, there are good arguments to revise the 2°C limit downward. Such a possible revision is actually foreseen in the Cancun Agreements, because the small island nations and least developed countries have long pushed for 1.5°C, for good reasons." Back at Cancun in 2010, the leading nations of the world agreed that they would "need to consider . . . strengthening the long-term global goal on the basis of the best available scientific knowledge, including in relation to a global average temperature rise of 1.5°C." Many studies had already come to the same conclusion. For instance the Royal Society's 2010 theme issue on 4°C warming noted, "the impacts associated with 2°C have been revised upwards, sufficiently so that 2°C now more appropriately represents the threshold between dangerous and extremely dangerous climate change."

Most recently, parties to the U.N. Framework Convention on Climate Change set up a "structured expert dialogue" from 2013 to 2015, to review the adequacy of the 2°C target. In May 2015, 70 of the world's leading climate experts who were involved in this dialogue reported back. They noted that, "Parties to the Convention agreed on an upper limit for global warming of 2°C, and science has provided a wealth of information to support the use of that goal." The authors state bluntly, "Limiting global warming to below 2°C necessitates a radical transition (deep decarbonization now and going forward), not merely a fine tuning of current trends." After reviewing the Fifth Assessment report and various presentations of observed climate impacts on regions around the world and agriculture, they point out, "Significant climate impacts are already occurring at the current level of global warming and additional magnitudes of warming will only increase the risk of severe, pervasive and irreversible impacts." As a result, they warn, "the 'guardrail' concept, which implies a warming limit that guarantees full protection from dangerous anthropogenic interference, no longer works." Here is their major conclusion:

> We are therefore of the view that Parties would profit from restating the long-term global goal as a 'defence line' or 'buffer zone', instead of a 'guardrail' up to which all would be safe. This new understanding would then probably favor emission pathways that will limit

warming to a range of temperatures below 2°C. In the very near term, such aspirations would keep open as long as possible the option of a warming limit of 1.5°C, and would avoid embarking on a pathway that unnecessarily excludes a warming limit below 2°C.

What kind of greenhouse gas emissions reductions are needed to achieve a 2°C target?

Peak global warming is primarily determined by cumulative (total) emissions of greenhouse gases. The longer the world delays global action, the more leaders will have to commit to deeper and faster emissions cuts. Roughly speaking, to have a significant chance—greater than 50%—of keeping total warming below 2°C, we need to cut the emissions of carbon dioxide and other major GHG pollutants by more than 50% by mid-century, which in turn means that global GHG emissions must peak within a decade or so and start a rapid decline. That decline must continue through century's end so that by 2100, the world's total net emissions of GHGs needs to be close to zero, and preferably below zero, especially if we delay serious action much longer. The goal is to stabilize atmospheric concentrations of CO_2 below 450 parts per million.

Such a target would still require that the sensitivity of the climate to carbon dioxide emissions over a multidecade period turns out not to be on the high end. And we will need to cut GHG emissions even more sharply than we thought if the amplifying carbon cycle feedbacks currently ignored by most global climate models start to kick in significantly in the next few decades.

In addition, the original U.N. Framework Convention on Climate Change—and subsequent protocols and agreements negotiated under its auspices—recognizes that as a matter of equity, some countries need to cut emissions faster than others. In particular, the developed nations that are industrialized and got "rich" by burning fossil fuels have been the biggest cumulative emitters of GHGs to date and have the highest emissions per capita. To be equitable, the countries with much greater wealth that resulted from much higher cumulative emissions have always been expected to cut GHG emissions considerably faster than the poorer countries, which are still developing. As a result, for most developed countries, an 80% to 90% reduction in GHGs by mid-century is the target needed to give the world a reasonable chance of stabilizing temperatures below 2°C.

What would the economic cost of meeting the 2°C target be?

Every major independent economic analysis of the cost of strong climate action has found that it is quite low.[41] In May 2014, the International Energy Agency (IEA) released its report on the cost of achieving the 2°C target, "Energy Technology Perspectives 2014." The IEA said that a systematic effort to use renewable energy and energy efficiency and energy storage to keep global warming below the 2°C threshold (their 2DS scenario) would require investment in clean energy of approximately 1% of global gross domestic product (GDP) per year. However, it would still be exceedingly cost-effective:

> The $44 trillion additional investment needed to decarbonise the energy system in line with the 2DS by 2050 is more than offset by over $115 trillion in fuel savings—resulting in net savings of $71 trillion.

A key point is that the investment is not the same as the net economic cost, because many of the investments reduce energy consumption and thus generate savings. In addition, investment in new technology is generally associated with higher productivity and economic growth.

The world's top scientists and economists made a similar finding in April 2014. That is when the U.N. Intergovernmental

Panel on Climate Change issued its Fifth Assessment report reviewing the scientific and economic literature on mitigation, which they define as "human intervention to reduce the sources or enhance the sinks of greenhouse gases." This assessment also looked at the cost of meeting the 2°C (3.6°F) target, a total greenhouse gas level in 2100 equivalent to 450 ppm of carbon dioxide. The IPCC determined that meeting such a target would reduce the median annual growth of consumption over this century by a mere 0.06%. In other words, the annual growth loss to avoid dangerous human-caused warming is 0.06%, and that is "relative to annualized consumption growth in the baseline that is between 1.6% and 3% per year."

In short, avoiding the worst climate impacts means global economic growth of some 2.24% a year rather than 2.30%. As always, every major government in the world signed off on every line of the report. This conclusion is not in much dispute (Table 4.1).

Table 4.1 Global mitigation costs for stabilization at a level "likely" to stay below 2°C (3.6°F). Cost estimates shown in this table do not consider the benefits of reduced climate change as well as co-benefits of mitigation. The three middle columns show the consumption loss in the years 2030, 2050, and 2100 relative to a baseline development without climate policy. The last column shows that the annualized consumption growth reduction over the century is 0.06%.

Consumption Losses in Cost-Effective Implementation Scenarios			
%Reduction in Consumption Relative to Baseline			Percentage Point Reduction in Annualized Consumption Growth Rate
2100 Concentration (ppm CO₂eq)	2030	2050	2100	2010–2100
450 (430–480)	1.7 (1.0–3.7) [N: 14]	3.4 (2.1–6.2)	4.8 (2.9–11.4)	0.06 (0.04–0.14)

Source: Intergovernmental Panel on Climate Change 2014.

Note that this cost estimate does not count the economic benefit of avoiding the most dangerous climate impacts. A few years ago, scientists calculated that benefit as having a net present value of as high as $830 *trillion*. These calculations do not include the so-called "co-benefits" of replacing relatively dirty fossil fuel-generated power with much cleaner sources of energy, including the use of more efficient technologies. Such co-benefits include reduced air pollution, improved public health, and the productivity gains associated with replacing old technology with new technology. An October 2014 IEA report concluded that "the uptake of economically viable energy efficiency investments has the potential to boost cumulative economic output through 2035 by USD 18 trillion." It specifically found that the co-benefits from energy efficiency upgrades alone equal, and often exceed, the energy savings.

The conclusion that avoiding dangerous warming has a very low net cost is not a new finding. In its previous Fourth Assessment in 2007, the IPCC found that the cost of stabilizing at the equivalent of 445 ppm carbon dioxide corresponded to "slowing average annual global GDP growth by less than 0.12 percentage points." These conclusions have remained consistent through time because they are based on a review of the literature, and every major independent study has found a remarkably low net cost for climate action and a high cost for delay.

For instance, in the private sector, the McKinsey Global Institute has done some of the most comprehensive and detailed cost analyses of how energy efficiency, renewable, and other low-carbon technologies could be used to cut GHG emissions. A 2008 McKinsey report, "The Carbon Productivity Challenge: Curbing Climate Change and Sustaining Economic Growth," concluded the following (emphasis added):

The macroeconomic costs of this carbon revolution are likely to be manageable, being in the order of 0.6–1.4 percent of global GDP by 2030. To put this figure in perspective, if one were to view this spending as a form of insurance against potential damage due to climate change, it might be relevant to compare it to global spending on insurance, which was 3.3 percent of GDP in 2005. Borrowing could potentially finance many of the costs, thereby effectively limiting the impact on near-term GDP growth. In fact, depending on how new low-carbon infrastructure is financed, **the transition to a low-carbon economy may increase annual GDP growth in many countries.**

As for the cost of delay, back in 2009, the IEA warned that "the world will have to spend an extra $500 billion to cut carbon emissions for each year it delays implementing a major assault on global warming." In its *World Energy Outlook 2011*, the IEA warned "Delaying action is a false economy: for every $1 of investment in cleaner technology that is avoided in the power sector before 2020, an additional $4.30 would need to be spent after 2020 to compensate for the increased emissions." The German economist Ottmar Edenhofer, who was the co-chair of the IPCC committee that wrote the 2014 report on mitigation, put it this way: "We cannot afford to lose another decade. If we lose another decade, it becomes extremely costly to achieve climate stabilization."

What happens if we miss the 2°C target?

The 2°C target is not a vertical cliff. It is more of a steep snowy slope down which we are pushing a rapidly growing snowball. At some point, the snowball will simply accelerate and expand on its own until it becomes a deadly avalanche. The Intergovernmental Panel on Climate Change says that the risks accumulate very quickly as we warm beyond 2°C. In May 2014, we learned that with the 0.85°C warming we have had to date, we are already at or close to the tipping point for the unstoppable collapse of a large part of the West Antarctic ice sheet. At the time, sea-level rise expert Stefan Rahmstorf pointed out, "More tipping points lie ahead of us. I think we should try hard to avoid crossing them."

In particular, there are some tipping points that also speed up the avalanche, the ones related to the carbon-cycle feedbacks. One of the biggest of those feedbacks, defrosting of the permafrost, is expected to start releasing carbon into the atmosphere in the 2020s, which, by itself, suggests we will ultimately have to cut greenhouse gases more sharply than currently anticipated. Similarly, the oceans and land-based carbon sinks (soils and vegetation) are becoming less efficient over time, which means nature will be providing us less and less help in removing carbon from the atmosphere over time. It is widely believed that the increasing inefficiency of the sinks is itself directly related to rising global temperatures. In some sense, we are in a race to see whether we can cut GHG emissions faster than the carbon-cycle feedbacks make that job more difficult.

Can we adapt to human-caused climate change?

Adaptation is how we deal with whatever climate change we are unable to prevent. In the case of sea-level rise, for instance, adaptation could include putting in place stilts on houses, along with levees, sea walls, pumping systems, and other engineering solutions to keep the rising water out. Otherwise, adaptation might simply be abandonment—leaving the inundated area. Generally richer communities will endeavor to find engineering adaptations to stay in place, whereas poorer ones will simply leave any region where the climate will no longer sustain life or livelihoods.[42]

The more climate change we allow by failing to aggressively reduce (mitigate) greenhouse gas emissions, the harder it will be to adapt in the sense of simply muddling through in place. "We basically have three choices: mitigation, adaptation and suffering," as Dr. John Holdren told the *New York Times* in

2007. The former president of the American Association for the Advancement of Science, whom President Obama appointed as national science advisor in 2009, said. "We're going to do some of each. The question is what the mix is going to be. The more mitigation we do, the less adaptation will be required and the less suffering there will be." The more global warming and climate change there is, the more limited the options become for adaptation. If sea levels were, say, 2 feet higher in 2100 and rising a few inches a decade, as was thought very possible until recently, one could envision many coastal communities adapting, albeit expensively. However, recent scientific findings suggest it will be more like, say, 6 feet higher in 2100 with sea level rising 1 foot a decade (or more) thereafter. Adapting to that is a considerably more expensive and difficult proposition. The *New York Times* story on the 2014 studies about the instability of the West Antarctic Ice sheet points out the risk posed by staying on our current GHG emissions path, given what we know: "The heat-trapping gases could destabilize other parts of Antarctica as well as the Greenland ice sheet, potentially causing enough sea-level rise that many of the world's coastal cities would eventually have to be abandoned."

It is extremely important to have as realistic an assessment of likely future impacts as possible in order to plan and prepare. In general, the risks of underestimating future impacts are considerably greater than the risks of overestimating them. That is because, as the Preface to Royal Society's special theme issue on a 4°C world explains, "responses that might be most appropriate for a 2°C world may be maladaptive in a +4°C world; this is, particularly, an issue for decisions with a long lifetime, which have to be made before there is greater clarity on the amount of climate change that will be experienced." The authors imagine a community building a reservoir to adapt to a moderate temperature increase, but that reservoir might simply go dry if the region gets very hot and dry. You might build a desalinization plant along your coast to provide water for a region that was drying out or losing a glacier-fueled river, but if you knew that rapid sea-level rise was coming, you would have to design an entirely different and probably much more expensive kind of facility to provide fresh water.

In the same way that an ounce of prevention is worth a pound of cure, mitigating future impacts through reductions in GHG emissions now is considerably cheaper than simply trying to adapt to high temperatures and rapid climate change in the future. A report on "Assessing the Costs of Adaptation to Climate Change" was published in 2009 by the International Institute for Environment and Development. It found that the mean "net present value of climate change impacts" in an emissions scenario similar to the one we are now on was $1240 trillion with no adaptation, but the value was only $890 trillion with adaptation. On the other hand, the authors report that in the "aggressive abatement" case (450 ppm), the mean "Net present value of climate change impacts" is only $410 trillion—or $275 trillion with adaptation. Stabilizing concentrations of CO_2 at 450 ppm reduces the net present value of impacts by $615 trillion to $830 trillion. However, the net present value of the abatement cost is only $110 trillion, a 6-to-1 savings for every dollar spent on cutting emissions. Therefore, although adaptation is certainly needed to reduce future impacts, abatement (mitigation) can reduce them far more.

In addition, some changes are very likely beyond our ability to deal with. The Intergovernmental Panel on Climate Change November 2014 "synthesis" of the scientific literature said we are risking "severe, pervasive and irreversible impacts for people and ecosystems." Scientists and governments have "high confidence" that these devastating impacts occur "even with adaptation," if we warm 4°C or more:

In most scenarios without additional mitigation efforts . . . warming is more likely than not to exceed 4°C [7°F] above pre-industrial levels by 2100. The risks associated with temperatures at or above 4°C include substantial species

extinction, global and regional food insecurity, consequential constraints on common human activities, and limited
potential for adaptation in some cases (*high confidence*).

Perhaps the hardest thing to adapt to besides rapid sea-level
rise is Dust-Bowlification. During the U.S. Dust-Bowl era in
the 1930s, some 3.5 million people fled the region of the Great
Plains. As I noted in "The Next Dust Bowl," a 2011 *Nature* article, "Human adaptation to prolonged, extreme drought is difficult or impossible," which is no doubt why the word "desert"
comes from the Latin for "an abandoned place."

It is not just extended drought that can dislocate people,
extreme heat with high humidity can also. In the business-as-
usual case, the IPCC warns that by 2100, "the combination of
high temperature and humidity in some areas for parts of the
year is expected to compromise common human activities,
including growing food and working outdoors (*high confidence*)." It is difficult to see how people, especially in the poorer
countries, could stay in a place that had essentially become
uninhabitable for parts of the year.

Here is one final example on "Agriculture and food systems in sub-Saharan Africa [SSA] in a 4°C+ world," from the
2010 Royal Society theme issue. That analysis concluded, "The
prognosis for agriculture and food security in SSA in a 4°C+
world is bleak." We already have nearly one billion people at
risk from hunger. In a 2°C world, it is estimated that achieving
food security would cost $40–$60 billion per year. However, as
we go past 2°C, the challenges go beyond simply money:

> Croppers and livestock keepers in SSA have in the past
> shown themselves to be highly adaptable to short- and
> long-term variations in climate, but the kind of changes
> that would occur in a 4°C+ world would be way beyond
> anything experienced in recent times. There are many
> options that could be effective in helping farmers adapt
> even to medium levels of warming, given substantial

investments in technologies, institution building and infrastructural development, for example, but it is not difficult to envisage a situation where the adaptive capacity and resilience of hundreds of millions of people in SSA could simply be overwhelmed by events.

Adaptation is sometimes called "resilience," which means the ability to bounce back. The best-case scenario, in which we keep total warming stabilized below 2°C, is one where genuine resilience is at least imaginable in many places. As we approach a business-as-usual level of warming, resilience is increasingly replaced by other forms of adaptation, including abandonment.

The 2015 Pacific Northwest National Laboratory study in *Nature Climate Change*, "Near-Term Acceleration in the Rate of Temperature Change," finds that by 2020, human-caused warming will move the Earth's climate system "into a regime in terms of multi-decadal rates of change that are unprecedented for at least the past 1,000 years." The rate of warming post-2050 becomes so fast that it is likely to be beyond adaptation (1) for most species and (2) for humans in many parts of the world. The warming rate hits 1°F per decade—Arctic warming would presumably be at least 2°F per decade, and this warming goes on for decades. Moreover, 4°C is not the worst-case scenario. If we go beyond 4°C, we move into an unrecognizable world where we will need a different word entirely than "adaptation."

What is geoengineering and can it play a major role in reducing the impact of climate change?

Geoengineering is not a well-defined scientific term. The broader definition is, "the large-scale manipulation of the Earth and its biosphere to attempt to counteract the effects of human-caused global warming." The term is so ill-defined and potentially misleading that the U.S. National Academy of

Sciences decided to reject the term "geoengineering" entirely and split its comprehensive peer-reviewed 2015 report on the subject into two: a 235-page report on "Carbon Dioxide Removal and Reliable Sequestration" and a 141-page report on "Reflecting Sunlight."

"Carbon dioxide removal" covers everything that could permanently take carbon dioxide out of the air—from reforestation to direct capture of carbon dioxide from the air. "Reflecting Sunlight," covers the more exotic climate-altering strategies to increase the reflectivity (albedo) of the Earth. The best studied of these is injecting vast quantities of sulfate aerosols into the stratosphere to mimic the cooling effect of volcanoes.

Instead of geoengineering, the Academy panel settled on "climate intervention," because "we felt 'engineering' implied a level of control that is illusory," explained Dr. Marcia McNutt (editor-in-chief of the journal *Science*) who led the report committee. The word "intervention" makes it clearer that the "precise outcome" could not be known in advance. Likewise, although many scientists have been referring to the reflecting sunlight strategies as "solar radiation management," the academy authors reject that in favor of "albedo modification" because, again, "management" implies a level of control of the outcome that the committee does not believe we have.

The basic conclusion of the report on carbon dioxide removal strategies is that they are relatively safe, but they are currently unaffordable and hard to scale up to the level needed to remove and dispose of billions of tons of carbon dioxide. The basic conclusion of the albedo modification report is that one or two of those strategies might be affordable, but they are dangerously flawed. Because of the problems with climate intervention, the central point the Academy makes in these reports is, "There is no substitute for dramatic reductions in the emissions of CO_2 and other greenhouse gases to mitigate the negative consequences of climate change, and concurrently to reduce ocean acidification." That is the same conclusion as a 2009 assessment by the UK Royal Society, which found "Geoengineering methods are not a substitute for climate change mitigation."

Capturing carbon dioxide from coal plants and storing it permanently is currently very expensive, as I will discuss in Chapter Six. However, the Carbon Dioxide Removal (CDR) strategies the Academy looked at are more expensive than extracting the carbon dioxide from coal burning (which does not count as CDR itself because you are not removing net carbon dioxide from the air with carbon capture and storage at a coal plant, you are just not adding new carbon dioxide into the air). As you might expect, direct air capture (i.e., simply pulling massive amounts of carbon dioxide out of thin air) is incredibly expensive and difficult to scale given that atmospheric carbon dioxide is so diffuse, only 400 parts per million. The Academy explains further:

> The barriers to deployment of CDR approaches are largely related to slow implementation, limited capacity, policy considerations, and high costs of presently available technologies. Additional research and analysis will provide information to help address those challenges. For these reasons, if carbon removal technologies are to be widely deployed, it is critical to embark now on a research program to lower the technical barriers to efficacy and affordability. In the end, any actions to decrease the excess burden of atmospheric CO_2 serve to decrease, or at least slow the onset of, the risks posed by climate change. Environmental risks vary among CDR approaches but are generally much lower than the risks associated with albedo modification approaches.

The scientific literature has repeatedly explained the limitations and risks of the aerosol-cooling strategy—or indeed any large-scale effort to manipulate sunlight. For one thing, they do nothing at all to slow the devastating impacts of ocean acidification discussed in Chapter Three (or any other impact that would be associated with rising carbon dioxide levels). One of the most widely discussed albedo modification strategies in the popular media is placing a significant number of reflectors or mirrors in outer space. However, the "the Committee has chosen to not consider these technologies because of the substantial time (>20 years), cost (trillions of dollars), and technology challenges associated with these issues."

In November 2014, the UK Guardian reported that the aerosol strategy "risks 'terrifying' consequences including droughts and conflicts," according to recent studies. "Billions of people would suffer worse floods and droughts if technology was used to block warming sunlight, the research found." Yet, stratospheric aerosol injection is considered the best albedo modification strategy.[43] The Academy concluded the following:

> Recommendation 3: Albedo modification at scales sufficient to alter climate should not be deployed at this time.
>
> There is significant potential for unanticipated, unmanageable, and regrettable consequences in multiple human dimensions from albedo modification at climate altering scales, including political, social, legal, economic, and ethical dimensions.

The Academy does endorse a program of mostly basic research into albedo modification in order to better "understand" it. That said, leading experts explained in the journal *Science* in 2010, "Stratospheric geoengineering cannot be tested in the atmosphere without full-scale implementation." In the article, "A Test for Geoengineering?", researchers explained that "weather and climate variability preclude observation of the climate response without a large, decade-long forcing. Such full-scale implementation could disrupt food production on a large scale"—for two billion people. In a December 2014 cover story, *Newsweek* interviewed Ken Caldeira, a well-known advocate for research into albedo modification and one of the Academy's panel members. They reported, "Caldeira doesn't believe any method of geoengineering is really a good solution to fighting climate change—we can't test them on a large scale, and implementing them blindly could be dangerous." The Academy report itself explained:

Albedo modification presents a number of risks and expected repercussions. Observed effects from volcanic eruptions include stratospheric ozone loss, changes to precipitation (both amounts and patterns), and likely increased growth rates of forests caused by an increase in diffuse solar radiation. Large volcanic eruptions are by their nature uncontrolled and short-lived, and have in rare cases led to widespread crop failure and famine (e.g., the Tambora eruption in 1815). However, effects of a sustained albedo modification by introduction of aerosol particles may differ substantially from effects of a brief volcanic eruption. Models also indicate that there would be consequences of concern, such as some ozone depletion or a reduction in global precipitation associated with sustained albedo modification. Further, albedo modification does nothing to reduce the build-up of atmospheric CO_2, which is already changing the make-up of terrestrial ecosystems and causing ocean acidification and associated impacts on oceanic ecosystems.

There is one final point on albedo modification. The Academy notes that "proposals to modify weather have tended to produce strong public opposition." Furthermore, there is "potential liability for any negative consequences" linked to a weather-modifying intervention. In particular, the Academy cites the case of "The first attempt to actually modify a hurricane," which "occurred in the late 1940s under Project Cirrus, a collaborative effort by the General Electric Company and the three military services." This was the seeding of an October 1947 hurricane off of the Florida-Georgia coast. Unexpectedly, "the seeded storm made an abrupt turn to the west and made landfall over the city of Savannah, Georgia." In this case, "Subsequent investigations and threats of litigation were successfully defended." However, "An important lesson is that those who conduct experiments that substantively alter weather—regardless of whether or not the interventions had any actual effect—can potentially be held legally liable for damage caused by the altered weather." Caldeira himself wrote me in 2011:

> Let's imagine the sunlight reflection method worked as advertised. How could anybody ever tell whether a weather event was due to the stratospheric aerosols, excess greenhouse gases, or natural variability in the climate system? If some region has a major drought in the decade after the introduction of the stratospheric aerosol spray, aren't they likely to attribute that change to the aerosol spray system? Isn't that likely to generate political friction and possibly even military conflict?

So beyond the liability issue—for example, a group who lost their farms in such a drought and was suing the group that did the aerosol injection—"even if the system worked as advertised, there is still great potential that socio-political risks could outweigh climate benefits." However, as Caldeira adds, "Of course, the system will not work as advertised." In that sense, geoengineering is akin to a risky, never tested, course of chemotherapy prescribed to treat a condition curable through diet and exercise—or, in this case, GHG emissions reduction.

Academy panel chair McNutt summed up their findings at the press conference, saying nobody should think that we can simply keep changing the climate with unrestricted carbon pollution and have any confidence we could intervene after the fact to fix things: "There is no silver bullet here, we cannot continue to release carbon dioxide and hope to clean it up later."

5

CLIMATE POLITICS
AND POLICIES

This chapter will explain the most commonly used or discussed climate policies around the world. It will also explore some of the issues involving climate politics.

What climate policies are governments around the world using to fight climate change?

The major policies used by governments to slow or reverse the growth in a country's greenhouse gas (GHG) emissions generally fall into four basic categories: economic, regulatory, technological, and forestry/land-use policies. The first category focuses on economic policies aimed at raising the price of carbon dioxide (and other GHG) emissions or subsidizing the cost of carbon-free energy sources. The goal of a carbon price is to have the economic cost of burning hydrocarbons (coal, oil, and natural gas) reflect the actual harm their emissions cause to humans and society. The two primary ways that carbon pricing is achieved is through a carbon tax or a cap-and-trade system, both of which are discussed below. Putting a price directly on CO_2 emissions does not necessarily mean that fossil fuels cannot be used to generate energy, only that doing so becomes more costly. The goal of subsidies for nuclear power or renewable forms of energy (such as solar and wind) is to similarly level the economic playing field but do so in a targeted manner that encourages deployment of new technologies, which generally brings their cost down.

The second category focuses on regulatory policies aimed at either increasing the use of clean energy or reducing the emissions of GHGs. Examples include fuel economy standards for vehicles, energy efficiency standards for appliances, renewable energy standards that require electricity (or vehicle fuel) to incorporate a certain minimum percentage of carbon-free sources, and limits on carbon dioxide emissions from different facilities such as electric power plants. The Clean Power Plan standards the U.S. Environmental Protection Agency (EPA) are pursuing is an example of a regulatory approach. Such policies can also target other major GHG emissions, such as methane and nitrous oxide.

The third category focuses on research-based policies aimed at lowering the cost and improving the performance of low-carbon sources. This includes basic research into new materials. It also includes applied research and development of advanced energy efficiency technologies such as LED lighting, a next-generation solar panel, and a lower-cost electric car battery. Research-based policy also includes helping to pay part of the cost needed to demonstrate the effectiveness of a low-carbon energy system operating on a large scale, such as a coal plant with carbon capture and storage.

The fourth category focuses on land and forestry policies aimed at reducing GHG emissions from deforestation and agricultural practices. Some countries, such as Brazil, have made significant reductions in their net emissions from deforestation. At one point, deforestation and land use policies were responsible for almost 20% of global GHG emissions, but now the number is closer to 10% if we just look at GHG emissions from deforestation and other land-use changes.

What is a carbon tax?

A carbon tax is a tax on the carbon content of hydrocarbon fuels or on the carbon dioxide emitted by those fuels when they are converted into energy. Hydrocarbons fuels—such as coal, oil, and natural gas—contain carbon, which turns into carbon dioxide after combustion. In economics, the total economic harm caused by a pollutant such as carbon dioxide can be considered an external cost that can be estimated and added to the price of that fossil fuel. If that "social cost of carbon" could fully account for all of the costs to society of emitting that pollutant, and if the tax were equal to that social cost, then businesses and other entities would reduce their use of fossil fuels in the most optimum and efficient manner. In practice, given the myriad projected impacts of climate change, many of which are unprecedented, coupled with uncertainties about exactly when these impacts will hit and how we should value future costs versus current ones, there is a large range in estimates of the social cost of carbon.

A number of countries have a carbon tax. Norway and Sweden introduced carbon taxes in 1991. Many other European countries also have a price on carbon content of fuel. In 2012, Australia introduced a $24 per metric ton carbon tax for major industrial emitters and some government entities. Much of the revenue raised was returned to the public in the form of lower income taxes or increased pensions and welfare payments. By mid-2014, the tax had cut carbon emissions by as much as 17 million metric tons, according to one study. The Australian government repealed the tax in July 2014.

In 2008, Canada's province of British Columbia (BC) launched the first economy-wide carbon tax in North America. It is "revenue neutral," which is to say that the revenues raised by the tax are returned to consumers and businesses in the form of lower personal and corporate taxes. If some of the revenues were used to pay for government spending, such as increased research and development into clean energy technologies, it would not be revenue neutral. The BC tax started at $10 per metric ton of carbon dioxide, and it hit $30 a metric ton in 2012. That translates into approximately $0.25 a gallon of gasoline. From 2008 to 2012, one study found that fossil fuel consumption fell 17% in BC (and 19% compared to the rest of Canada).[44]

Many countries that do not have a significant (or any) carbon price do place a large tax on petroleum-based fuels, such as gasoline and diesel. These taxes are often substantially larger on gasoline than a typical carbon tax would be, but they pay for road repair and offset other externality costs associated with fuel consumption. In many European countries and Japan, the gasoline tax is typically a few dollars a gallon, whereas the vast majority of carbon taxes in place today add a cost to the price of gasoline that is one tenth that size.

What are cap-and-trade and carbon trading?

Cap-and-trade is a market-based environmental policy aimed at reducing pollution. It has emerged as one of the most popular means of reducing greenhouse gases (and other pollutants) worldwide. The European Union (EU) Emission Trading System is a cap-and-trade system for carbon dioxide adopted in 2003 that is "by far the world's largest environmental pricing regime."[45]

In a cap-and-trade system, the "cap" is a limit set by a governmental entity (such as the EPA) on the amount of a specific pollutant (such as carbon pollution or acid rain pollution) that an entire industrial sector (such as the utility industry) can emit. The cap is enforced by having that entity allocate or sell a limited number of permits or allowances, which give companies the "right" to pollute a certain amount. Companies cannot emit a pollutant without allowances to do so.

Those permits can then be sold on secondary markets comparable to the stock exchange. Companies that can cheaply and efficiently reduce their emissions below their allocation can sell those permits to other companies that find reducing emissions more costly. This is the "trade" part of cap and trade. The allocated permits are reduced over time in specified fashion, which reduces the overall level of pollution. A shrinking cap generally means a rising price for the pollutant being traded. By specifying in advance how the cap or allocation will be reduced, this policy sends a long-term signal to businesses and other players in the marketplace that the price of this pollutant is likely to rise over time, thus incentivizing long-term investment in technologies and strategies that can reduce or replace CO_2.

A cap-and-trade system is similar to a carbon tax in that both aim to set a price for carbon dioxide that will reduce its emission into the atmosphere. However, the cap-and-trade system lets the market set the price for carbon dioxide, whereas in a tax, the government sets the price. In theory, the cap-and-trade system is considered to be more flexible, economically efficient, and business friendly than the so-called "command-and-control" regulations made popular in the 1970s. Those regulations typically required a company to make a specific reduction in air or water pollution at every single facility it owned, even if some facilities or some companies could easily and cheaply make far deeper reductions and other facilities or companies could not. The cap-and-trade system is designed to achieve a comparable target level of overall economy-wide emissions reductions as the other pollution-reduction strategies while (1) rewarding the companies that are the most innovative or efficient at cutting pollution and (2) making certain that the target level of emissions is achieved at the least possible cost.

A 2013 article in the *Journal of Economic Perspectives* by two leading economic experts on cap-and-trade explain some of its history:

> In the 1980s, President Ronald Reagan's Environmental Protection Agency put in place a trading program to phase out leaded gasoline. It produced a more rapid elimination of leaded gasoline from the marketplace than had been anticipated, and at a savings of some $250 million per year compared with a conventional no-trade, command-and-control approach. Not only did President George H. W. Bush successfully propose the use of cap-and-trade to cut US SO_2 [sulfur dioxide] emissions,

his administration advocated in international forums the use of emissions trading to cut global CO_2 emissions, a proposal initially resisted but ultimately adopted by the European Union. In 2005, President George W. Bush's EPA issued the Clean Air Interstate Rule, aimed at reducing SO_2 emissions by a further 70 percent from their 2003 levels. Cap-and-trade was again the policy instrument of choice.

The sulfur dioxide (or acid rain pollution) trading program was a cornerstone of the Clean Air Act Amendments of 1990, which passed the U.S. Senate by a vote of 89 to 11 and the House of Representatives by a vote of 401 to 21, including 87% of Republican members and more than 90% of Democrats in both houses. The trading system helped industry beat the emissions reduction targets at a lower cost than anyone had projected. An EPA study ultimately found that the 1990 Clean Air Act Amendments had "benefits exceeding costs by a ratio of 25-to-1" by 2010. Costs in 2010 were approximately $53 billion, whereas benefits were $1.3 trillion. By 2010, the cumulative benefits included 1.8 million lives saved, 1.3 million heart attacks prevented, 137 million additional days of work (and productivity) because workers were healthier, and 26 million more school days because students were also healthier.

It was, in part, the economic, environmental, and political success of the acid rain program that led the Europeans to embrace cap and trade for greenhouse gases. The EU enacted their Emission Trading System to meet the targets they had committed to under the 1997 Kyoto protocol, which was to reduce collective EU emissions to 8% below 1990 levels during the 5-year period between 2008 and 2012. The EU Emission Trading System was criticized for issuing too many allowances, which ultimately led to a very low trading price for carbon dioxide, "less than 4 euros (around $5.25) per ton of carbon, down from nearly 30 euros in 2008," as the *New York Times* explained in 2013. However, thanks in large part to the trading system, "Emissions have fallen by 14 percent among sectors covered by the program in countries that have participated since 2005." So it was successful in achieving its goal and is likely to be central to the EU meeting the carbon dioxide target their members announced in 2014—cutting GHG emissions 40% below 1990 levels by 2030.

Likewise, the success of the acid rain program inspired many states to embrace a cap-and-trade system. For instance, California put in place a cap-and-trade system in 2013 to help meet its target of returning GHG emissions to 1990 levels by the year 2020—with the eventual goal of an 80% reduction from 1990 levels by 2050 for the state. In 2003, George Pataki, then Republican Governor of New York, reached out to the governors of Mid-Atlantic and Northeastern states "to develop a strategy that will help the region lead the nation in the effort to fight global climate change." That in turn led to the creation of the Regional Greenhouse Gas Initiative in 2008, a cap-and-trade program whose goal was to cut CO_2 emissions in the utility sector 10% by 2018. At one point, it encompassed 10 states: Connecticut, Delaware, Maine, Maryland, Massachusetts, New Hampshire, New Jersey, New York, Rhode Island, and Vermont. New Jersey exited RGGI in 2011.

There have been many efforts to start a nationwide cap-and-trade system for carbon dioxide in the United States. Senator John McCain, a Republican from Arizona, repeatedly introduced legislation in the United States Senate to set up such a system, but he never garnered sufficient votes to pass it. In 2009, the U.S. House of Representatives did pass a cap-and-trade bill, but legislation was ultimately never taken up in the Senate. Criticisms of the bill included concerns that it would be too complex and costly.

Many other countries are initiating cap-and-trade programs. In 2011, China launched pilot carbon trading in several cities and provinces, including Beijing, Shanghai, and Shenzhen along with Guangdong and Hubei Provinces. China has subsequently announced that it will launch its national carbon market in 2016. In the fall of 2014, China pledged to cap total carbon dioxide emissions by 2030 and, the Chinese cap-and-trade system is widely expected to become the largest carbon market in the world. In January 2015, South Korea launched a program as part of its effort to cut GHG emissions to 30% below current levels by 2020. South Korea's carbon market is already one of the largest in the world.

What is China doing to restrict carbon dioxide emissions?

In November 2014, Chinese President Xi Jinping joined President Obama in the U.S.-China Joint Announcement, which stated "China intends to achieve the peaking of CO_2 emissions around 2030 and to make best efforts to peak early." This was the first time a major developing nation—in this case, the largest and fastest-growing carbon-emitter in the world—agreed to sharply change the trajectory of its carbon emissions and fossil fuel consumption. This joint announcement was widely viewed as increasing the chances for a successful global treaty in Paris in December 2015. China experts I spoke to believe that China would not have agreed to the phrase "to make best efforts to peak early," if their leaders did not believe that they could and would do so.[46]

The Chinese had already started to work toward a CO_2 peak, with energy price reforms, strong fuel economy standards, and an aggressive effort to deploy clean energy technologies, which had made them world leaders in both manufacturing and utilizing solar power and wind power. In the new announcement, China committed to "increase the share of non-fossil fuels in primary energy consumption to around 20% by 2030." That clean energy pledge "will require China to deploy an additional 800-1,000 gigawatts of nuclear, wind, solar, and other zero emission generation capacity by 2030," as the White House explained at the time. And that is "more than all the coal-fired power plants that exist in China today and close to total current electricity generation capacity in the United States."

One week later, the Chinese government announced it would cap coal use by 2020. The Chinese State Council, or cabinet, said the peak would be 4.2 billion metric tons, a one-sixth increase over current consumption. This was a major reversal of Chinese energy policy, which for 2 decades had been centered on building a coal plant or more a week. Now, they will be building the equivalent in carbon-free power every week for decades, while the construction rate of new coal plants slows sharply.

A key motivation for China to cut coal use, beyond simply its interest in slowing climate change, is that their urban air pollution levels are among the highest in the world. This gives them a major public health and domestic political motivation to peak coal soon. To meet its CO_2 and air pollution targets, for instance, Beijing province needs to slash coal use by 99% by 2030, to fewer than 200,000 metric tons. Su Ming, a leading Chinese energy expert working at an institute run by China's National Development and Reform Commission explained to Reuters, "We are trying to tell provincial officials how much coal they could use under a restricted nationwide quota." That would mean "the big consuming regions of Hebei, Tianjin and Shandong" would have to cut coal use by up to 27% by 2030.

In February 2014, China reported it had cut its coal consumption 2.9% in 2014, the first drop this century. Domestic coal production fell 2.5%. Melanie Hart, the Director for China Policy at the Washington, DC think tank the Center for American Progress (where I work), explained to me that the 2014 coal drop suggests the country is "well on track to peak coal use by 2020", but "if coal growth remains sluggish over the next few years Chinese coal use could possibly peak even earlier." Also, she points out that "most models indicate that China's carbon dioxide emissions will peak about ten years after coal."

I visited China at the end of June 2015 to meet with top governmental and non-governmental experts on clean energy and climate. That visit made clear to me that the country's leaders are serious about reversing their energy policy and cleaning up their polluted air. That is a key reason the Chinese are beating climate and clean energy targets across the board. China may well have already peaked or at least plateaued in coal consumption back in 2013. It's now a widely held view in the Beijing climate community that China will peak its carbon dioxide emissions by 2025.

What is the United States doing to restrict carbon dioxide emissions?

In the 2014 U.S.-China Joint Announcement on climate, President Obama advanced "a new target to cut net greenhouse gas emissions 26-28% below 2005 levels by 2025." Meeting that target roughly doubles the rate of decline Obama had committed the United States to with his previous target of a 17% cut by 2020, which had been announced in the weeks leading up to the 2009 Copenhagen climate summit.

The United States has embraced a variety of strategies at the federal and state level to cut carbon pollution. In the last few years, the United States adopted much-strengthened fuel economy standards for automobiles for the first time in decades. Half the states have renewable electricity standards that require their electric utilities to purchase a significant fraction of their power from new renewable power, such as solar and wind. The federal government has had a long-term program of investing in renewable energy research, development, demonstration, and deployment, which has helped bring down the cost of renewable power. Some states have specific programs to cut carbon pollution, such as the Regional Greenhouse Gas Initiative used by several northeastern and Mid-Atlantic States, discussed earlier. California has enacted very strong GHG regulations that require steady and deep reductions in its CO_2 emissions, ultimately leading to an 80% cut in emissions below 1990 levels by 2050.

These strategies, coupled with the lower cost of natural gas (made possible by the fracking revolution) along with aggressive state and federal and corporate efforts to increase energy efficiency, have reversed the steady rise of U.S. GHG emissions. As of 2014, U.S. carbon dioxide emissions were 8% below 2005 levels.

The primary additional strategy the United States has embraced to meet its target in 2020 and beyond is the use of the Clean Air Act, which requires the EPA to regulate pollutants found to endanger the public. In 2007, the U.S. Supreme Court ruled in Massachusetts v. EPA that carbon dioxide does qualify as a pollutant under the Clean Air Act. In 2009, the EPA found that based on the science, CO_2 was a danger to public health. That meant the EPA had to put in place CO_2 standards for "mobile sources," which it did when the new U.S. fuel economy standards went into effect. After that, the EPA was required to put in place CO_2 standards for stationary sources, including the biggest sources, which are new and existing electric power plants.

In 2014, the Supreme Court affirmed 7 to 2 that indeed the EPA has the authority to regulate GHGs from stationary sources, such as power plants.[47] The EPA has already put forward regulations on new power plants and, as of mid-2015, was finalizing such regulations for existing power plants. These regulations would give each state a target for how much CO_2 can be emitted per unit of electricity generated. The states can then devise for themselves the specific plan for its electricity providers to achieve that target, which could include building new renewable and nuclear power plants, replacing existing coal plants with new natural gas plants, running cleaner plants more often (and dirty plants less often) than they had been, embracing energy efficiency programs, attempting to capture and store carbon dioxide from power plants, or even putting in place a carbon tax or a cap-and-trade system.

How do different political parties view climate science and policies in the United States and around the world?

In most countries around the world, opposing political parties tend to share the overwhelming scientific consensus that climate change is a serious problem that must be addressed. The European Union for instance, has been embracing stronger and stronger greenhouse gas targets for 2 decades—and putting in place stronger and stronger policies to achieve those targets—even as many of their governments have flipped from conservative to liberal or vice versa. In the case of the United Kingdom, one UK newspaper noted back in 2010 that "All the major parties are signed up to transforming Britain into a green, low-carbon economy to boost growth, as well as to combat climate change."[48]

In 2009, when some UK climate scientists were being attacked, both the Labour government led by Prime Minister Gordon Brown and leaders of the Conservative Party reiterated support for climate science:

> But tonight the shadow climate change secretary, Greg Clark, made clear the party line remains that climate change is a serious man-made threat. "Research into climate change has involved thousands of different scientists, pursuing many separate lines of independent inquiry over many years. The case for a global deal is still strong and in many aspects, such as the daily destruction of the Earth's rainforests, desperately urgent," he said.

In the UK's 2015 parliamentary election, all parties reiterated their support for Britain's climate targets, although they do differ on specific policies of how to get to that target.

A handful of countries in the world are politically divided on climate change. For instance, Australia's Labor government introduced a carbon tax in 2012. In the federal election of 2013, the opposition Liberal Party campaigned against the tax and after winning the election, it was able to repeal the tax in 2014.

The United States is notably divided on climate change at the level of national politics, with the national Democratic Party and its leaders generally in favor of strong action on climate change and the national Republican Party and its leaders now generally opposed. However, the partisan divide was not always so sharp. As recently as 2008, both the Democratic nominee for president (Senator Barack Obama) and the Republican nominee for president (Senator John McCain) were both running on a platform of putting in place a cap-and-trade system for the United States. In 2008, many leading national Republicans endorsed climate action. Also, at the state level, many Republican governors have been advocates of strong climate action. It was Republican Governor George Pataki of New York who launched the effort that resulted in the Regional Greenhouse Gas Initiative, which put in place a cap-and-trade system for several Northeastern and Mid-Atlantic States. Republican governor Arnold Schwarzenegger of California championed that state's strong carbon dioxide targets, and he remains a strong advocate of climate action.

At the national level, however, the Republican Party has become strongly opposed to climate action, with many elected leaders expressing doubt or disbelief of basic climate science. In 2010, the *National Journal* reported that "The GOP is stampeding toward an absolutist rejection of climate science that appears unmatched among major political parties around the globe, even conservative ones." The magazine pointed out that in contrast, British Foreign Secretary William Hague along with "such other prominent European conservatives as French President Nicolas Sarkozy and German Chancellor Angela Merkel have embraced" the widespread scientific consensus on climate change and "supported vigorous action."

More recently, Senate Majority Leader Mitch McConnell, a Republican from Kentucky, started an "aggressive campaign to block" EPA's carbon pollution standards "in statehouses and courtrooms across the country, arenas far beyond Mr. McConnell's official reach and authority," as the *New York Times* reported. In March 2015, McConnell sent a letter to every governor in the country urging him or her not to comply with federal law. McConnell is trying to "undercut Mr. Obama's position internationally as he tries to negotiate a global climate change treaty to be signed in Paris in December," the *Times* reported. "The idea is to create uncertainty in the minds of other world leaders as to whether the United States can follow through on its pledges to cut emissions."

McConnell asserts that state inaction "will provide time for the courts to rule on whether the EPA's proposed rule is legal." As discussed above, the EPA is legally obligated to issue rules regulating CO_2 from existing power plants, and, in 2014, the Supreme Court reaffirmed 7 to 2 that the EPA has that authority. Under U.S. law, if states do not put forward a state implementation plan to meet the EPA's CO_2 standards, the federal government is required by law to do so. Unless the Supreme Court reverses itself, states can delay their reductions through legal action, but ultimately they will have to make them.

Is there a large-scale effort to spread misinformation on climate science and, if so, who funds it?

For more than 2 decades, the fossil fuel industry has been funding scientists, think tanks and others to deny and cast doubt on the scientific understanding of human-caused global warming. The major media has extensively documented this. As recently as February 2015, a *New York Times* exposé revealed that a researcher at the Harvard-Smithsonian Center for Astrophysics who routinely casts doubt on widely accepted climate science "has accepted more than $1.2 million in money from the fossil-fuel industry over the last decade while failing to disclose that conflict of interest in most of his scientific papers." This included funding from Exxon-Mobil and "at least $230,000 from the Charles G. Koch Charitable Foundation." In books and documentaries such as "Merchants of Doubt: How a Handful of Scientists Obscured the Truth on Issues from Tobacco Smoke to Global Warming," historians and journalists have shown (1) that this misinformation and disinformation campaign goes all the way back to the tobacco industry's campaign to cast doubt on claims that cigarette smoking is bad for your health and (2) that in some cases it involves the same exact people.[49]

In 2009, the *New York Times* documented that the Global Climate Coalition, an anti-action lobbying group backed by industries that profit from fossil fuels, ignored its own climate scientists during the 1990s while spreading disinformation about global warming.[50] An internal report stating that the human causes of global warming "cannot be denied" fell on the deaf ears of Coalition leaders. The Coalition led an "aggressive lobbying and public relations campaign against the idea that emissions of heat-trapping gases could lead to global warming." However, the final draft of a 1995 "Primer on Climate Change Science" written by the GGC's own scientific experts—which was made public years later through a federal lawsuit—revealed that those experts "were advising that the science backing the role of greenhouse gases in global warming could not be refuted." For instance, those experts concluded the following: "The scientific basis for the Greenhouse Effect and the potential impact of human emissions of greenhouse gases such as CO_2 on climate is well established and cannot be denied." In addition, after a long analysis of "Are There Alternate Explanations for the Climate Change Which Has Occurred Over the Last 120 Years?" they conclude: "The contrarian theories raise interesting questions about our total understanding of climate processes, but they do not offer convincing arguments against the conventional model of greenhouse gas emission-induced climate change."

The *New York Times* reported that the Global Climate Coalition "was financed by fees from large corporations and trade groups representing the oil, coal and auto industries, among others." They had a substantial budget:

In 1997, the year an international climate agreement that came to be known as the Kyoto Protocol was negotiated, its budget totaled $1.68 million, according to tax records obtained by environmental groups.

Throughout the 1990s, when the coalition conducted a multimillion-dollar advertising campaign challenging the merits of an international agreement, policy makers and pundits were fiercely debating whether humans could dangerously warm the planet.

Ultimately, the *Times* notes, "The coalition, according to other documents, later requested that the section of the primer endorsing the basics of global warming science be cut." In the same way that the tobacco industry knew of the dangers of smoking and the addictive nature of nicotine for decades, but their CEOs and representatives publicly denied those facts, many of those denying the reality of human plus climate science have long known the actual science. In July 2015, we learned that oil giant Exxon understood the scientific reality of climate change as far back as 1981, many years before climate change became a political issue that they tried to spread confusion about.

Over the years, fossil fuel companies and their executives were documented to have funneled tens of millions of dollars into this disinformation campaign. For a long time, the leading funder was the oil company Exxon-Mobil. However, they have been overtaken by Koch Industries—a company with large fossil fuel interests, run by billionaires Charles and David Koch—which spent more $48.5 million from 1997 to 2010 to fund disinformation. A report concluded that "From 2005 to 2008, Exxon Mobil spent $8.9 million while the Koch Industries-controlled foundations contributed $24.9 million in funding to organizations of the climate denial machine."

In 2015, the *Times* revealed that Dr. Wei-Hock "Willie" Soon had taken more than $1 million dollars from Exxon-Mobil, the Kochs, and other fossil fuel interests without generally disclosing that conflict of interest in his scientific papers. During this period, Soon has advanced a repeatedly debunked theory arguing that humans are not the primary cause of global warming. As the *Times* explained:

> Though he has little formal training in climatology, Dr. Soon has for years published papers trying to show that variations in the sun's energy can explain most recent global warming. His thesis is that human activity has played a relatively small role in causing climate change.

The *Times* goes on to explain "Many experts in the field say that Dr. Soon uses out-of-date data, publishes spurious correlations between solar output and climate indicators, and does not take account of the evidence implicating emissions from human behavior in climate change." The head of NASA's Goddard Institute for Space Studies explained that solar variability probably is responsible for at most 10% of recent global warming, whereas human-caused GHGs are responsible for the overwhelming majority of it. He added, "The science that Willie Soon does is almost pointless."

In October 2014, the Smithsonian itself put out a climate statement, which makes clear that such a view is simply anti-scientific. The Smithsonian explains, "Scientific evidence has demonstrated that the global climate is warming as a result of increasing levels of atmospheric greenhouse gases generated by human activities." The newly uncovered documents show that "Dr. Soon, in correspondence with his corporate funders, described many of his scientific papers as 'deliverables' that he completed in exchange for their money." The Smithsonian repeatedly signed off on contracts with Southern Company Services—a coal company and long-time funder of science denial—requiring the Smithsonian to provide the coal utility "advanced written copy of proposed publications . . . for comment and input."

The fossil fuel industry has known for 2 decades that the solar variability explanation for recent climate change is untrue. As far back as 1995, the scientific and technical advisors to the Global Climate Coalition wrote in their draft primer:

> [The] hypothesis about the role of solar variability and [Pat] Michaels' questions about the temperature record are not convincing arguments against any conclusion that we are currently experiencing warming as the result of greenhouse gas emissions. However, neither solar variability nor anomalies in the temperature record offer a mechanism for off-setting the much larger rise in temperature which might occur if the atmospheric concentration of greenhouse gases were to double or quadruple.

We have been headed for a tripling of atmospheric concentrations of carbon dioxide, and the dangerous consequences of doing so are widely understood and accepted by the world's leading climate scientists and governments. The multidecade disinformation campaign funded by the fossil fuel industry is marked by a rejection of basic science and the constant repetition of flawed arguments that have been long debunked by scientists, even ones who were advising the fossil fuel industry. That disinformation campaign continues today with more money than ever.

What are climate science deniers?

The scientific community and leading governments of the world have repeatedly reported on the ever-strengthening body of research supporting our understanding of basic climate science. Some people, including a small number of scientists who do research in the climate arena, reject this science. Those people are often called "climate science deniers" or climate deniers, especially those who receive financial support from fossil fuel interests. Some who reject climate science embrace the term "denier," whereas others reject it, preferring the term "skeptic." All scientists, however, are skeptics, and many believe that "denier" is a more accurate term for those who reject climate science.[51]

"Based on well-established evidence, about 97% of climate scientists have concluded that human-caused climate change is happening," explained The American Association for the Advancement of Science explained in its 2014 report, "What We Know." The world's largest general scientific society explained:

> The science linking human activities to climate change is analogous to the science linking smoking to lung and cardiovascular diseases. Physicians, cardiovascular scientists, public health experts and others all agree smoking causes cancer. And this consensus among the health community has convinced most Americans that the health risks from smoking are real. A similar consensus now exists among climate scientists, a consensus that maintains climate change is happening, and human activity is the cause.

The media does not write about "tobacco science skeptics" and no longer gives airtime to people who deny the dangerous health consequences of cigarette smoking. However, many in the media continue to quote those who deny basic climate science. In December 2014, four dozen leading scientists and science journalists/communicators issued a statement urging the media to "Please stop using the word 'skeptic' to describe deniers" of climate science. The 48 signatories from the United States, the United Kingdom, and around the world are Fellows of the Committee for Skeptical Inquiry. They include Nobel laureate Sir Harold Kyoto; Douglas Hofstadter, Director of The Center for Research on Concepts and Cognition at Indiana University; physicist Lawrence Krauss, Director of The Arizona State University Origins Project; and Bill Nye "the Science Guy."

The scientists and journalists were motivated by a November 2014, *New York Times* article, "Republicans Vow to Fight EPA and Approve Keystone Pipeline" that referred to Senator James Inhofe (R-OK) as "a prominent skeptic of climate change." They note that in the same week, National Public Radio's Morning Edition called Inhofe "one of the leading climate change deniers in Congress." The signatories note, "These are not equivalent statements" and the two terms should not be conflated.

"Proper skepticism promotes scientific inquiry, critical investigation, and the use of reason in examining controversial and extraordinary claims," the letter reads. "It is foundational to the scientific method. Denial, on the other hand, is the a priori rejection of ideas without objective consideration." The scientists and journalists point out that Inhofe's assertion that global warming is "the greatest hoax ever perpetrated on the American people" is a very extraordinary claim of a "vast alleged conspiracy." They note that true skepticism is embodied in a quote often repeated by Carl Sagan: "extraordinary claims require extraordinary evidence." However, the Senator has never been able to provide even ordinary evidence for his conspiracy charge. "That alone should disqualify him [Inhofe] from using the title 'skeptic'."

The signatories explain that they are "skeptics who have devoted much of our careers to practicing and promoting scientific skepticism." They ask journalists to "stop using the word 'skeptic' to describe deniers"—those who reject basic climate science. They write

> As scientific skeptics, we are well aware of political efforts to undermine climate science by those who deny reality but do not engage in scientific research or consider evidence that their deeply held opinions are wrong. The most appropriate word to describe the behavior of those individuals is "denial." Not all individuals who call themselves climate change skeptics are deniers. But virtually all deniers have falsely branded themselves as

skeptics. By perpetrating this misnomer, journalists have granted undeserved credibility to those who reject science and scientific inquiry.

Some of the people labeled "deniers" take offense at the apparent implication that they are like Holocaust deniers. Some people have tried to coin other terms, such as "denialist" or "disinformer." However, coining terms is nearly impossible, and "deniers" remains a term that is widely embraced, including by many deniers themselves. As the National Center for Science Education explained in their 2012 post, "Why Is It Called Denial?"

> "Denial" is the term preferred even by many deniers. **"I actually like 'denier.' That's closer than skeptic,"** says MIT's Richard Lindzen, one of the most prominent deniers. Minnesotans for Global Warming and other major denier groups go so far as to sing, "I'm a Denier!"

Thus, using the term denier does not inherently mean you are equating a disinformer with a Holocaust denier. Moreover, the overwhelming majority of people who use the term certainly do not mean it in that sense. That said, people who use the term would be well advised to explain what they do and do not mean by it.

6

THE ROLE OF CLEAN ENERGY

This chapter will focus on the energy technologies most widely discussed for a transition to a low carbon economy. It will explore the scale of the energy transition needed to explain why some energy technologies are considered likely to be major contributors to the solution and others not.

What kind of changes in our energy system would a 2°C target require?

To have a significant chance of keeping total warming below 2°C, we need to cut global emissions of carbon dioxide and other major greenhouse gas (GHG) pollutants by more than 50% by mid-century. That rapid decline needs to continue through 2100, by which time the world's total net emissions of greenhouse gases should be close to zero, if not below zero.

In its 2014 Fifth Assessment report reviewing the scientific and economic literature on mitigation, the United Nations Intergovernmental Panel on Climate Change reported that "CO_2 emissions from fossil fuel combustion and industrial processes contributed about 78% of the total GHG emission increase from 1970 to 2010, with a similar percentage contribution for the period 2000–2010." That is why policymakers focus so much attention on our energy system.

Avoiding 2°C warming means deep cuts in emissions from fossil fuel combustion—except in the scenario where carbon capture and storage (CCS) from fossil fuel plants is commercially feasible on a wide scale, the prospects of which are discussed later in this chapter. However, CCS is likely to play a limited role in the next few decades, as a 2015 article in the journal *Nature* explained: "Because of the expense of CCS, its relatively late date of introduction (2025), and the assumed maximum rate at which it can be built, CCS has a relatively modest effect on the overall levels of fossil fuel that can be produced before 2050 in a two-degree scenario." This article, "The Geographical Distribution of Fossil Fuels Unused When Limiting Global Warming to 2 °C," concluded the following: "Our results suggest that, globally, a third of oil reserves, half of gas reserves and over 80% of current coal reserves should remain unused from 2010 to 2050 in order to meet the target of 2 °C."

The International Energy Agency (IEA) is one of the few independent organizations in the world with a sophisticated enough global energy model to analyze in detail what different emissions pathways would mean for the global energy system. One IEA report, their "World Energy Outlook 2011", "presents a 450 Scenario, which traces an energy path consistent with meeting the globally agreed goal of limiting the temperature rise to 2°C." They note that eventually, all scenarios become "locked-in by existing capital stock, including power stations, buildings and factories." The IEA concludes: "Without further action by 2017, the energy-related infrastructure then in place would generate all the CO_2 emissions allowed in the 450 Scenario up to 2035." In other words, the world cannot afford to build much new fossil fuel infrastructure. New infrastructure needs to be primarily carbon free. Any new fossil fuel energy system needs to be accompanied by shutting down equivalent old ones. In addition to that, we need to start replacing existing fossil fuel energy systems with carbon-free ones.

The core energy-related climate solutions are those that can supply vast amounts of carbon-free energy or that can use vast amounts of energy much more efficiently in the next few decades. On the energy supply side, the core climate solutions are nuclear power, renewable energy such as wind and solar power, and potentially CCS from fossil fuel power plants. This chapter will examine the supply options with the most promise for displacing a significant amount of current CO_2-emitting energy sources by mid-century. It will also look at opportunities to reduce energy demand through energy efficiency and conservation. Finally, it will look at opportunities to reduce GHG emissions in the agricultural sector.

What is energy efficiency and what role will it play?

Energy efficiency is reducing the energy consumption of our products and services, while maintaining or improving their performance. Efficiency is the most important climate solution for several reasons:

1. It is by far the biggest resource.
2. It is by far the cheapest, far cheaper than the current cost of unsustainable energy, so cheap that it helps pay for the other solutions.
3. It is by far the fastest to deploy, without the transmission and siting issues that plague most other strategies.
4. It is "renewable"—the efficiency potential never runs out.

Energy efficiency can be as simple as insulating a home to reduce the energy consumed in heating and cooling. Energy efficiency in transportation could mean improving the fuel economy of vehicles. Efficiency in lighting could be replacing inefficient incandescent light bulbs with light-emitting diode (LED) bulbs, and it could also be using occupancy sensors to turn off the lights automatically when everyone has left the room. Moreover, it could be designing your building to make greater use of daylight, so less electrical lighting is needed in the first place.

In November 2014, the International Energy Agency released its "Energy Efficiency Market Report 2014." The report found "The global energy efficiency market is worth at least USD 310 billion a year and growing." The IEA's Executive Director Maria van der Hoeven summarized the core findings: "Energy efficiency is the invisible powerhouse in IEA countries and beyond, working behind the scenes to improve our energy security, lower our energy bills and move us closer to reaching our climate goals." The IEA explains that "in the IEA scenario consistent with limiting the long-term increase in global temperatures to no more than 2 degrees Celsius, the biggest share of emissions reductions—40%—comes from energy efficiency." The report itself provides data "confirming energy efficiency's place as the *first fuel*."

In 15 years of experience working with businesses—at the U.S. Department of Energy (DOE) and at nonprofit think tanks and a consulting firm—I have generally found that any home, commercial building, or manufacturing facility could cut energy consumption and carbon dioxide emissions by 25% to 50% or more while reducing its energy bill and increasing productivity. This could be initiated with a return on investment that generally exceeded 25% and in many cases 50% to 100%. In 1999, I published the first collection of detailed case studies, some 100 in all, of how businesses were cutting energy use and boosting productivity while reducing pollution—*Cool Companies: How the Best Businesses Boost Profits and Productivity by Cutting Greenhouse Gas Emissions*. Five years before that, green design guru Bill Browning and I published, "Greening the Building and the Bottom Line: Increasing Productivity Through Energy-Efficient Design," a Rocky Mountain Institute report peer-reviewed by the U.S. Green Building Council. However, those case studies were often dismissed as mere anecdotes, even though many subsequent reports by energy efficiency and design experts supported their findings.

In October 2014, however, the IEA, the global body responsible for energy analysis, also reported that the non-energy benefits from energy efficiency upgrades equal (and often exceed) the energy savings. The 232-page report upends decades of conventional thinking about efficiency.

The most noteworthy conclusion of "Capturing the Multiple Benefits of Energy Efficiency" may be that "the uptake of economically viable energy efficiency investments has the potential to boost cumulative economic output through 2035 by USD 18 trillion," which is larger than the current size of the U.S. economy.

In particular, the report finds that green building design can achieve health benefits, including reduced medical costs and higher worker productivity, "representing up to 75% of overall benefits." That is, the nonenergy benefits of energy efficiency upgrades can be three times the size of the energy savings. The IEA study also finds that when the value of productivity and operational benefits of industrial efficiency measures were factored into "traditional internal rate of return calculations, the payback period for energy efficiency measures dropped from 4.2 to 1.9 years." Payback time was cut in half.

The report's core finding on buildings, if widely embraced by developers, architects, building owners, and governments, could revolutionize both building design and public policy (emphasis added):

> Energy efficiency retrofits in buildings (e.g. insulation retrofits and weatherisation programmes) create conditions that support improved occupant health and well-being, particularly among vulnerable groups such as children, the elderly and those with pre-existing illnesses. The potential benefits include improved physical health such as reduced symptoms of respiratory and cardiovascular conditions, rheumatism, arthritis and allergies, as well as fewer injuries. **Several studies that quantified total outcomes found benefit-cost ratios as high as 4:1 when health and well-being impacts were included, with health benefits representing up to 75% of overall benefits**. Improved mental health (reduced chronic stress and depression) has, in some cases, been seen to represent as much as half of total health benefits.

The IEA reports that if you include lower total public health spending, then "Addressing indoor air quality through energy efficiency measures could, in a high energy efficiency scenario, save the European Union's economy as much as $259 billion annually." On the industrial side, the IEA concludes, "The value of the productivity and operational benefits derived can be up to 2.5 times (250%) the value of energy savings (depending on the value and context of the investment)."

These results are consistent with many private sector analyses. For instance, the global consulting company McKinsey has repeatedly documented how an aggressive energy efficiency strategy sharply lowers the cost of climate action. In 2009, they released their most comprehensive analysis to date of the U.S. energy efficiency opportunity, "Unlocking Energy Efficiency in the U.S. Economy." They concluded that a comprehensive set of efficiency measures in the residential, commercial, and the industrial sector, if fully enacted over the next decade, would save 1.2 billion tons of carbon dioxide equivalent, which was 17% of U.S. CO_2-equivalent emissions in 2005. McKinsey explained the "central conclusion of our work":

> Energy efficiency offers a vast, low-cost energy resource for the U.S. economy—but only if the nation can craft a comprehensive and innovative approach to unlock it. Significant and persistent barriers will need to be addressed at multiple levels to stimulate demand for energy efficiency and manage its delivery across more than 100 million buildings and literally billions of devices. If executed at scale, a holistic approach would yield gross energy savings worth more than $1.2 trillion, well above the $520 billion needed through 2020 for upfront investment in efficiency measures (not including program costs). Such a program is estimated to reduce end-use energy consumption in 2020 by 9.1 quadrillion BTUs, roughly 23 percent of projected demand, potentially abating up to 1.1 gigatons of greenhouse gases annually.

*Will nuclear power be a major factor in the effort
to minimize climate change?*

"Nuclear energy remains the largest source of low-carbon
electricity" in the developed countries, providing 18% of their
power, noted a 2015 report from the International Energy
Agency and Nuclear Energy Agency (NEA). The report also
explained that nuclear was the second-biggest source of
low-carbon electricity in the world, with an 11% share.

Because it is a low-carbon source of around-the-clock (base-
load) power, a number of climate scientists and others have
called for a re-examination of nuclear policy. The Chinese in
particular have been building nuclear power plants at a steady
pace. However, very few new plants have been ordered and
built in the past 2 decades in countries with market economies,
such as the United States, which derives 20% of its power from
nuclear. That is primarily because new nuclear plants are so
costly, but also because dealing with the radioactive nuclear
waste remains problematic and the costs of an accident are so
enormous.

In particular, the Fukushima nuclear disaster in Japan
caused a number of countries, including Japan and Germany,
to reconsider their dependence on nuclear power. On March
11, 2011, the Fukushima Daiichi nuclear power plant north of
Tokyo was hit by a wall of water 43 feet high, which was triggered by a massive earthquake. This destroyed or disabled
enough equipment to cause three reactors to partially melt
down. It also caused a loss of water at the open-air ponds that
store radioactive spent nuclear fuel rods. In 2014, Japanese
college professors calculated that the accident "will cost 11.08
trillion yen ($105 billion), twice as much as Japanese authori-
ties predicted at the end of 2011." That includes both radiation
clean up and compensation paid to the victims.[52]

The costs of new nuclear reactors have been rising for
decades, and they are now extremely expensive, costing up to
$10 billion dollars apiece. A key reason new reactors are inher-
ently so expensive is that they must be designed to survive

almost any imaginable risk, including major disasters and human error. Even the most unlikely threats must be planned for and eliminated when the possible result of a disaster is the poisoning of thousands of people, the long-term contamination of large areas of land, and $100 billion in damages.

Since Fukushima, global nuclear capacity growth, which was not very fast to begin with, slowed considerably. In 2014, there were only three new plants under construction—and just 5 gigawatts (GW) of capacity were added. In their 2015 report titled "Technology Roadmap: Nuclear Energy," the IEA and Nuclear Energy Agency explain what level of capacity additions would be required in its 2 degrees Celsius (2D) scenario: "In order for nuclear to reach its deployment targets under the 2D scenario, annual connection rates should increase from 5 GW in 2014 to well over 20 GW during the coming decade." That means returning to a nuclear build rate previously achieved for only 1 decade, i.e., 20 GW/year during the 1980s. That target has many challenges in a post-Fukushima world. The IEA and NEA themselves note that "Such rapid growth will only be possible" if several actions take place including, "Vendors must demonstrate the ability to build on time and to budget, and to reduce the costs of new designs." If such advances do occur, then new nuclear power could provide 5% to 10% of the needed new carbon-free power for the 2°C scenario.

For the medium term, the DOE and others have been working to develop small modular reactors that could start to ramp up production in 2030 and beyond. These reactors are constructed in factories and would cost $3–$5 billion each. Ideally, they would be much safer than the large reactors. However, because they are smaller and generate much less electricity, it is not clear that their cost per kilowatt-hour (kWh) of delivering electricity would be much lower than current nuclear plants.

In the longer term, if nuclear power is going to continue to play a major role through mid-century and beyond, the excessive water consumption of typical nuclear reactors is something that will have to be addressed in a world where a considerable fraction of the habited land of the planet is severely drying out. A typical nuclear reactor today uses 35–65 million liters of water each day. Two plants in Georgia use more water than all the water used by people living in Atlanta, Augusta, and Savannah combined.

What role does natural gas have in the transition to a 2°C world?

Natural gas plays a big role in powering and heating the global economy. Because natural gas is the least carbon-intensive of all the fossil fuels and because it can be burned much more efficiently than either coal or oil, we are likely to continue burning a considerable amount of natural gas for the next few decades. However, a number of studies, including comprehensive surveys and analyses by the U.N. Intergovernmental Panel on Climate Change and the International Energy Agency, suggest that if we are to stabilize global temperatures below 2°C, natural gas has a very short window in which its use can increase, before, like coal and oil, it peaks and begins a rapid decline that ends with little natural gas being used by century's end.[53]

Natural gas has two big challenges as it seeks to play a bigger role in the global energy economy. First, it is a hydrocarbon, and when it is burned, it releases carbon dioxide into the atmosphere. Under some circumstances, natural gas power plants can replace coal-fired power plants and achieve significant reduction in CO_2 emissions, as much as 50%. However, whenever natural gas displaces new or existing nuclear power, renewable energy, or energy efficiency, then it represents an increase in CO_2 emissions.

In 2013, Stanford's Energy Modeling Forum published the results of economic modeling by more than a dozen different expert teams. This modeling found that more abundant, cheaper shale gas (extracted using fracking technology) has little impact on annual growth in United States' greenhouse gas emissions through 2050 compared with the case of low shale gas. Why? Over time, and especially post-2020, "natural gas begins to displace nuclear and renewable energy that would have been used otherwise in new power plants under reference case conditions." Cheap natural gas can also slow the shift toward more efficient use of energy. Likewise, a 2014 study, "The Effect Of Natural Gas Supply on US Renewable Energy and CO_2 Emissions," confirmed that "increased natural gas use for electricity will not substantially reduce US GHG emissions, and by delaying deployment of renewable energy technologies, may actually exacerbate the climate change problem in the long term." Over time, and especially post-2020, "natural gas begins to displace nuclear and renewable energy that would have been used otherwise in new power plants under reference case conditions."

The second challenge natural gas faces in a carbon-constrained world is that natural gas is mostly methane (CH_4), a super-potent GHG, which traps 86 times as much heat as CO_2 over a 20-year period. So even small leaks in the natural gas production and delivery system can have a large near-term climate impact—enough to eliminate the entire benefit of switching from coal-fired power to gas for many decades. The actual leakage rate from natural gas production and delivery is not known with certainty, but several major studies in the past few years have found that it is high enough to neutralize the benefits of a coal to gas switch for an extended period of time. For instance, a 2014 study in the journal *Science*, "Methane Leaks from North American Natural Gas Systems," reviewed more than 200 earlier studies. It concluded that natural gas leakage rates were approximately 5.4%. At that leakage rate, replacing a fleet of coal plants with natural gas plants would have no climate benefit for 50 years. That is, replacing coal plants with gas plants would be worse for the climate for some 5 decades. Later in 2014, satellite observations of huge oil and gas basins in East Texas and North Dakota found very high leakage rates of heat-trapping methane. "In conclusion," researchers write, "at the current methane loss rates, a net climate benefit on all time frames owing to tapping unconventional resources in the analyzed tight formations is unlikely."

The methane leakage issue is not settled, and it is possible that aggressive regulation by governments can induce industry to cut the leakage rate measurably. However, for now, it is difficult to see how a significant and sustained increase in natural gas use is consistent with the 2°C scenario. In addition, as noted above, because natural gas does not just displace coal but also displaces carbon-free forms of power, even studies that do not factor in a significant leakage rate find little if any climate benefit from major investment in new natural gas infrastructure.

In 2011, the IEA concluded in its "World Energy Outlook 2011" report, "Without further action by 2017, the energy-related infrastructure then in place would generate all the CO_2 emissions allowed in the 450 Scenario up to 2035." As discussed above, this means the world cannot afford to build much new fossil fuel infrastructure. Rather, new infrastructure needs to be primarily carbon free. In its "Energy Technology Perspectives 2012" report, another major IEA study found very little room for growth in a 2°C scenario: "The specific emissions from a gas-fired power plant will be higher than average global CO_2 intensity in electricity generation by 2025, raising questions around the long-term viability of some gas infrastructure investment if climate change objectives are to be met." By 2030, natural gas is increasingly playing a supporting role to renewables: "Post-2030, as CO_2 reductions deepen in the 2DS, gas-powered generation increasingly takes the role of providing the flexibility to complement variable renewable energies and serves as peak-load power to balance generation and demand fluctuations."

The IPCC came to a similar conclusion in their 2014 review of mitigation strategies, which include both natural gas and renewable energy. They found, "In the majority of low stabilization scenarios, the share of low carbon electricity supply increases from the current share of approximately 30%

to more than 80% by 2050." That kind of rapid growth in near-zero-carbon energy over the next 35 years leaves very little room for any new fossil fuel generation. The IPCC asserts that natural gas can act as a short-term bridge fuel if "the fugitive emissions associated with extraction and supply are low or mitigated." Currently, however, it does not seem as if fugitive emissions are low, nor do we know if they can be reduced sufficiently and, if so, whether governments and industry will take the steps needed to do so.

How much can solar power contribute to averting dangerous climate change?

Solar power is the use of the sun's energy to generate electricity. "The sun could be the world's largest source of electricity by 2050, ahead of fossil fuels, wind, hydro and nuclear," according to two 2014 reports from the International Energy Agency. There are two major types of solar power: Photovoltaics (PV), direct conversion of sunlight into electricity, and concentrated solar thermal power, which uses the heat from focused sunlight to generate electricity. In recent years, PV has vastly surpassed solar thermal power in terms of both cost and demand. This trend seems likely to continue for the next decade or two, and solar PV is poised to become one of the pillars of a low-carbon economy in the coming decades (along with wind). However, studies suggest that solar thermal power may regain its competitive advantage after 2030. The two IEA *Technology Roadmap* reports show that PV and solar thermal power combined could generate as much as 27% of the world electricity by mid-century.[54]

Some materials generate an electric current or voltage when exposed to photons of light, hence the photovoltaic effect. The first photovoltaic cell dates back to 1839, but Bell Laboratories in the United States invented the first practical solar cell in 1954 using the semiconducting material silicon. These first cells, expensive and inefficient, were only suitable for niche applications, such as outer space, where traditional forms of power are impractical. Eventually, as the global semiconductor industry advanced, increasing performance while cutting costs, the solar industry was able to piggyback on its advances.

The price of solar photovoltaic cells dropped 99% from 1977 to 2013, and it has continued dropping since then. From 2008 to 2014 alone, solar prices have dropped 80%. In an increasing number of markets around the world, the cost of electricity provided by onsite solar power is now at or very close to the cost of power from the electric grid. As a result, solar PV manufacturing capacity and sales have been growing rapidly. A 2013 study from Stanford University noted, "If current rapid growth rates persist, by 2020 about 10% of the world's electricity could be produced by PV systems." Commitments by China, the European Union, and other countries to dramatically expand renewable power to reduce carbon dioxide emissions make it very likely that rapid growth rates will persist.

From the perspective of CO_2 mitigation, the energy needed to make a PV systems, its energy intensity, has also been steadily declining. The Stanford study found that "if the energy intensity of PV systems continues to drop at its current learning rate, then by 2020 less than 2% of global electricity will be needed to sustain growth of the industry," even while solar power is providing 10% of global electricity. That means that solar power can be a sustainable source of very low carbon power in the coming decades.

It is ironic, as one of the Stanford authors noted, "At the moment, Germany makes up about 40% of the installed market, but sunshine in Germany isn't that great. So from a system perspective, it may be better to deploy PV systems where there is more sunshine." Germany has done the world a great favor by investing so heavily in its renewable energy transition, which has helped to bring down the cost of solar energy for every country. However, solar power is considerably more cost-effective in places where it is sunnier longer during both the day and year, such as the Southwest United States and the Middle East. This is another reason we can expect the amount of solar PV generated to continue its rapid increase.

Concentrated solar thermal power (CSP) is the use of mirrors to focus sunlight to heat a fluid that runs an engine to make electricity. The first commercial CSP plant dates back to 1913, when a 55-kilowatt CSP water-pumping station using parabolic mirrors was installed in Egypt. Two of the basic CSP designs used today are mirrors that focus the light on a long tube and the power tower, which uses many mirrors moving in two dimensions to focus sunlight on a central tower that holds the engine. The key attribute of CSP is that it generates primary energy in the form of heat, which can be stored 20 to 100 times more cheaply than electricity, with far greater efficiency. Commercial projects have already demonstrated that CSP systems can store energy by heating oil or molten salt, which can retain the heat for hours.

Both the U.S. National Renewable Energy Laboratory and the IEA have said that after solar photovoltaics makes a deep penetration into the electricity market, CSP may well become more valuable. That is because right now, solar PV generates electricity at the most valuable time—daily peak electricity usage in daytime, particularly during the summertime, when air conditioning becomes the big draw on electric power. However, once solar PV hits 10% to 15% of annual electric generation in a region, PV can become less valuable. The IEA projects that when that occurs, perhaps around 2030, "Massive-scale [solar thermal electric] deployment takes off at this stage thanks to CSP plants' built-in thermal storage, which allows for generation of electricity when demand peaks in late afternoon and in the evening, thus complementing PV generation."

The IEA 2014 roadmaps show that with the right policies and continued technology improvement, PV could provide as much as 16% of the world's electricity by 2050 while CSP plants could provide 11%. In that scenario "Combined, these solar technologies could prevent the emission of more than 6 billion tonnes of carbon dioxide per year by 2050—that is more than all current energy-related CO_2 emissions from the United States or almost all of the direct emissions from the transport sector worldwide today."

How big a role will wind power play in averting dangerous climate change?

Wind power uses the power of the blowing wind to turn blades. That rotation can provide the energy to directly spin a wheel or a gear, which can do anything from turn a wheel that grinds grain to spin a turbine generator that makes electricity. Wind power has been the fastest-growing form of new renewable power in terms of added electricity generation per year. By the end of 2014, some 370 gigawatts of wind power were installed in the world, with nearly 52 GW installed in 2014 alone. The International Energy Agency noted in its 2013 *Technology Roadmap: Wind Power* that "In a few countries, wind power already provides 15% to 30% of total electricity." With continued improvement in the technology, the IEA projects that as much as 18% of the world's electricity could be provided by wind power in 2050. [55]

More than 2000 years ago, simple windmills were used in the Middle East to grind grain, and in China they were used to pump water. Windmills were introduced to Europe in the 11th century, and by the 18th century, the Netherlands alone was using more than 10,000 windmills. The mills were ultimately replaced by steam engines because they could not compete with the low-cost and reliability of fossil fuels. Wind power began to see a resurgence in the 1970s because of the energy crises that raised the price of fossil fuels. In the past few decades, significant aerodynamic improvements in blade design and other advances have brought down the cost of electricity from wind power by 10% per year. Wind energy can now be captured efficiently over broad ranges of wind speed and direction, so turbines do not have to be located in the windiest regions to be economical.

Such advances have helped give wind power production an annual growth rate averaging more than 20%. This has led to economies of scale in production, which, combined with larger turbine blades and improved control systems, have kept the price drops continuing. As the IEA puts it: "The technology keeps rapidly improving, and costs of generation from land-based wind installations have continued to fall. Wind power is now being deployed in countries with good resources without special financial incentives."

Wind power is easily the cheapest new form of electricity in Denmark, according to a 2014 government analysis. The Danish Energy Agency projects that wind plants coming online in 2016 will provide power for 5.4 cents per kilowatt-hour, half the price of new coal and natural gas plants. In 2013, Denmark generated one third of its electricity from wind power, and in December 2013, for the first time, wind power provided more than half of Denmark's electricity consumption.

The 2014 Stanford study on energy payback times found that the wind industry has been in energy surplus for decades. That is, the wind industry is generating far more energy than it took to build all of the world's wind turbines. Today, more than 90% of the electrical output of the onshore wind industry is now available to society. So wind power can continue its rapid growth rate and remain a sustainable source of low carbon power.

What is carbon capture and storage (a.k.a. carbon sequestration) and what role can it play?

Burning fossil fuels releases carbon dioxide into the atmosphere, and that carbon dioxide is the primary cause of recent global warming. In general, the vast majority of strategies to reduce such carbon dioxide emissions involve reducing fossil fuel combustion, either by replacing fossil fuels with carbon-free sources (such as solar energy or nuclear power) or by using a technology that does the same job but simply uses less energy (such as an energy efficient light bulb or motor). Carbon sequestration, also known as carbon capture and storage (CCS), is a technology that is being pursued which might allow the continued use of fossil fuels, especially coal. Unfortunately, CCS has developed more slowly than expected, and the technology is unlikely to make a major contribution to reducing carbon pollution until after the 2020s.[56]

To ensure fossil fuel combustion does not release carbon pollution into the atmosphere, the carbon dioxide from a coal-fired power plant (or potentially a gas-fired one) must be captured and stored somewhere forever. That carbon dioxide could be removed before combustion or after combustion. Doing so before you burn the fossil fuel is much simpler and cheaper because after combustion, the carbon dioxide begins to diffuse in the exhaust (flue) gas and then the atmosphere. The more diffuse the carbon dioxide, the more difficult and costly it is to extract from the air.

On the other hand, coal can be gasified, and the resulting "synthesis gas" (syngas) can be chemically processed to produce a hydrogen-rich gas and a concentrated stream of carbon dioxide. The latter can be piped directly to a carbon storage site. The former can be burned in a highly efficient "combined cycle" power plant. The whole process—integrated gasification combined cycle (IGCC) plus permanent storage in underground sites—is considerably more expensive than conventional coal plants. In 2009, Harvard's Belfer Center for Science and International Affairs published a major study, "Realistic Costs of Carbon Capture." The Harvard analysis concluded that first-of-a-kind CCS plants will have a cost of carbon abatement of some "$150/tCO2 [$150 per ton of carbon dioxide" avoided, not counting transport and storage costs. This yields a "cost of electricity on a 2008 basis [that] is approximately 10 cents/kWh higher with capture than for conventional plants." That price would effectively double the cost of power from a new coal plant. In 2003, The National Coal Council explained a key problem that is slowing development of IGCC: "Vendors currently do not have adequate economic incentive" to pursue the technology because "IGCC may only become broadly competitive with" under a "CO_2-restricted scenario."

It would certainly be more useful to have a CCS technology that could capture and store the carbon dioxide from the exhaust or flue gas postcombustion produced by *thousands* of existing coal plants than to have CCS technology that works only on newly designed plants. However, that technology has historically been even further from commercialization at scale and necessarily involves capturing carbon dioxide that is far more dilute. As a 2008 U.S. DOE report had pointed out:

> "Existing CO_2 capture technologies are not cost-effective when considered in the context of large power plants. Economic studies indicate that carbon capture will add over 30% to the cost of electricity for new integrated gasification combined cycle (IGCC) units and over 80% to the cost of electricity if retrofitted to existing pulverised coal (PC) units. In addition, the net electricity produced from existing plants would be significantly reduced—often referred to as parasitic loss—since 20-30% of the power generated by the plant would have to be used to capture and compress the CO_2."

Given how very expensive early-stage carbon capture and storage is, jump-starting accelerated development and deployment of CCS requires:

1. A rising price on carbon dioxide to make CCS profitable or
2. Large subsidies by some government entity or
3. Significant investment and financing by the private sector or
4. Some combination of those three things

Perhaps the major reasons for the very slow development of CCS for both new and existing power plants have been (1) lack of a price on carbon dioxide or other government policy that could provide large ongoing subsidies coupled with (2) lackluster interest and investment by the private sector.

How slow has development been? In October 2013, the *New York Times* summarized the state of CCS with their headline, "Study Finds Setbacks in Carbon Capture Projects." The story noted that, "the technology for capturing carbon has not been proved to work on a commercial scale, either in the United States or abroad." One major CCS demonstration at a West Virginia coal plant was shut down in 2011 because "it could not sell the carbon dioxide or recover the extra cost from its electricity customers, and the equipment consumed so much energy that, at full scale, the project would have sharply cut electricity production."

The 2013 survey on the "Global Status of CCS," by the Global CCS Institute found that "while C.C.S. projects are progressing, the pace is well below the level required for C.C.S. to make a substantial contribution to climate change mitigation." The 2014 survey by the Institute found continued progress, but it concluded, "the data on large-scale CCS projects highlights two other areas requiring increased attention by policymakers—the lack of projects in non-OECD economies (outside of China) and the lack of progress in CCS technology development in high carbon intensive industries such as cement, iron and steel and chemicals." The Norwegian oil and gas company Statoil is one of the few in the world that has actually captured carbon dioxide from natural gas processing facilities and stored it geologically (in former gas and oil fields). Statoil's vice president for CCS said in late 2014, "Today the cost per ton is economically prohibitive," and so "We need public-private partnerships where the government takes commercial exposure and some of the risks."

A key issue for CCS is that although the development of large-scale commercial projects has been very slow, the requirements for CCS to make a major dent in the global warming problem are huge. Vaclav Smil, Distinguished Professor Emeritus of the Environment at the University of Manitoba in Canada, described "the daunting scale of the challenge," in his analysis "Energy at the Crossroads":

> "Sequestering a mere 1/10 of today's global CO_2 emissions (less than 3 Gt CO_2) would thus call for putting in place an industry that would have to force underground every year the volume of compressed gas larger than or (with higher compression) equal to the volume of crude oil extracted globally by [the] petroleum industry whose infrastructures and capacities have been put in place over a century of development. Needless to say, such a technical feat could not be accomplished within a single generation."

There are many other issues with CCS. For instance, there is the leakage issue. Even a very small leakage rate from an underground carbon storage side (less than 1% a year) would render it all but useless as a "permanent repository." In addition, a Duke University study found the following: "Leaks from carbon dioxide injected deep underground to help fight climate change could bubble up into drinking water aquifers near the surface, driving up levels of contaminants in the water tenfold or more in some places." What kind of contaminants could bubble up into drinking water aquifers? The study noted: "Potentially dangerous uranium and barium increased throughout the entire experiment in some samples." This problem may not turn out to be fatal to CCS, but it might well limit the places where sequestration is practical, either because the geology of the storage site is problematic or because the site is simply too close to the water supply of a large population.

Public acceptance has already been a major problem for CCS. Public concern about CO_2 leaks (small and large) has impeded a number of CCS projects around the world. Modest leaks risk water contamination, but large leaks can actually prove fatal because in high concentrations, CO_2 can suffocate people. As *BusinessWeek* reported in 2008:

> "One large, coal-fired plant generates the equivalent of 3 billion barrels of CO_2 over a 60-year lifetime. That would require a space the size of a major oil field to contain. The pressure could cause leaks or earthquakes, says Curt M. White, who ran the US Energy Department's carbon sequestration group until 2005 and served as an adviser until earlier this year. 'Red flags should be going up everywhere when you talk about this amount of liquid being put underground.'"

With the use of hydraulic fracturing to produce natural gas in the United States, we have seen considerable concern about leakage of methane and other potentially harmful substances. There is a growing body of research linking hydraulic fracturing to earthquakes. That has been especially true for the so-called reinjection wells, where millions of gallons of wastewater from the fracturing process are injected deep underground, much as the carbon dioxide would be in CCS. Research published by Stanford University concluded in 2012:

> We argue here that there is a high probability that earthquakes will be triggered by injection of large volumes of CO_2 into the brittle rocks commonly found in continental interiors. Because even small- to moderate-sized earthquakes threaten the seal integrity of CO_2 repositories, in this context, large-scale CCS is a risky, and likely unsuccessful, strategy for significantly reducing greenhouse gas emissions.

Concern about leaks helped kill one of the world's first full CCS demonstrations of capturing, transporting, and storing carbon dioxide by the Swedish company Vattenfall in northern Germany. The project started operation in 2008. In 2009, Germany tried to introduce legislation that would have had the government assume liability for companies injecting carbon dioxide underground. The legislation failed to pass. Vattenfall did not get a permit to bury the carbon dioxide. As a result in July 2009, the plan "ended with CO_2 being pumped directly into the atmosphere, following local opposition at it being stored underground." In May 2014, the company announced that it was ending all CCS research.

Carbon capture and storage has a long way to go to become a major contributor to addressing the threat of climate change starting in the 2030s. We will need vastly more effort by the public and private sector if CCS is going to provide as much as 10% of the carbon dioxide emissions reductions needed by 2050.

What is bioenergy and what is its role in cutting carbon pollution?

Human beings have used biomass (plant matter) for cooking and heating since the discovery of fire. In many rural parts of the world, fuel wood is still the main source of energy for cooking and heating. Many companies have a low-cost biomass waste stream that they use to generate power. Today, biomass power generation is a relatively small and slow-growing fraction of overall electricity generation. However, biofuels (such as corn ethanol) have become a significant and fast-growing source of liquid fuels for transportation. Many of the most popular of the current generation biofuels are made from food crops, and whether they provide any overall greenhouse gas benefit is disputed in the literature. [57]

The future of bioenergy will depend critically on the availability of biomass for energy use. Biomass is a very inefficient means of turning sunlight into energy. For instance, research has found that corn ethanol from Iowa converts only 0.3% of incoming solar radiation into sugar and only some 0.15% into ethanol. In addition, transporting biomass long distances has a high monetary and energy cost. Finally, internal combustion engines are a very inefficient means of energy conversion. All that inefficiency and energy loss means that you need a huge amount of land to deliver low-carbon biofuels to the wheels of a car, especially compared with say solar and wind power used to charge an electric vehicle. Considerable agricultural land is used today to grow bioenergy crops, such as corn ethanol and palm oil. However, if the world does not quickly get on the 2°C path, then temperature rise, sea-level rise, ocean acidification, and Dust-Bowlification combined make it unlikely there will be enough arable land and potable water to feed 10 billion people post-2050 *and* grow substantial energy crops—unless major advances in next-generation biofuels occur.

The current generation of crop-based biofuels is, with one major exception, very problematic as a carbon-reducing strategy. Many biofuels, including biodiesel from rapeseed and biofuels from palm oil, have been found to have higher total GHG emissions than the petroleum fuels they replace, when a full lifecycle analysis is done. In Indonesia, burning forests to clear land for palm oil production has became a major source of carbon emissions.

Corn ethanol—the biofuel most widely used in the United States—has been widely criticized in the scientific literature for directly and indirectly consuming a great deal of land, which in turn means it generates GHG emissions comparable to that of the gasoline it replaces. An April 2015 article in *Environmental Research Letters*, "Cropland Expansion Outpaces Agricultural and Biofuel Policies in the United States," is the "first comprehensive analysis of land-use change across the United States between 2008 and 2012." University of Wisconsin-Madison researchers "tracked crop-specific expansion pathways across the conterminous US and identified the types, amount, and locations of all land converted to and from cropland" during that time. Scientists learned that crops "expanded onto 7 million acres of new land," during those four years and replaced "millions of acres of grasslands." Half of that was new soy and corn, which was increasingly used to make biofuels during that time to meet U.S. government mandates, which included a minimum target of over 12 billion gallons (45 billion liters) of biofuels in 2010.

What was the climate impact? The University of Wisconsin-Madison concluded, "The conversion to corn and soy alone, the researchers say, could have emitted as much carbon dioxide into the atmosphere as 34 coal-fired power plants operating for one year—the equivalent of 28 million more cars on the road." It remains far from clear whether corn ethanol has any net climate benefit.

The one crop-based biofuel that appears to have a clear climate benefit is ethanol made from sugarcane, which is typically grown in tropical climates. Brazil in particular has devoted significant resources to sugarcane ethanol. In 2010, the U.S. Environmental Protection Agency certified that ethanol from sugarcane cuts greenhouse gases 61% compared to gasoline, based on a detailed lifecycle analysis. Sugarcane produces twice as much ethanol per acre of land as corn does. The process also produces a waste-byproduct, bagasse, which can be used to generate power for the operation. Sugarcane also stores a great deal of carbon both above ground and in soils. In addition, it only requires replanting every 6 years, which means less land tilling and likely less carbon released from the soil.

The problem with relying on crop-based biofuels, as noted, is that we already have nearly 1 billion people globally suffering from starvation. We are going to add 3 billion more people over the next few decades—all of whom will need food and water—while we turn as much as one third of the arable land into Dust Bowls with near-permanent drought. At the same time, sea-level rise and ocean acidification will further constrain agriculture and food production.

That is why scientists and governments have been working on a second-generation biofuel. So-called cellulosic ethanol is not made from the starchy edible parts of plants but rather the entire plant, including the parts we do not eat. In theory, it can be made from forestry and agricultural waste (corncobs, stalks, husks) as well as dedicated energy crops, such as prairie grass and switchgrass. After many decades of research on the problem, three commercial cellulosic ethanol plants finally opened in the United States in 2014. However, even now, the question of whether all forms of cellulosic ethanol, such as the nonedible parts of corn, are actually major carbon reducers remains open.

If we do get on the 2°C pathway, and if we continue to see technology advances, then it is plausible that cellulosic bioenergy could play a role. In 2011, the International Energy Agency's report "Technology Roadmap: Biofuels for Transport" noted that "Biofuels provide only around 2% of total transport fuel today." They projected a best-case scenario that "By 2050, biofuels could provide 27% of total transport fuel and contribute in particular to the replacement of diesel, kerosene and jet fuel." If indeed electricity from carbon-free sources becomes a predominant transport fuel for personal vehicles, then the greatest need for low-carbon transport fuels will be for other uses such as air travel and trucking.

In 2012, the IEA's "Technology Roadmap: Bioenergy for Heat and Power", projected a best-case scenario for 2050 whereby bioenergy provides 7.5% of world electricity generation plus 15% of heat needed in industry and 20% of heat in the buildings sector. That would represent 2 billion tons of CO_2-equivalent emissions savings, but it can only happen "if the feedstock can be produced sustainably and used efficiently, with very low life-cycle GHG emissions." If carbon capture and storage ever proves feasible on a large scale, then using it with sustainable low-carbon bioenergy would actually create a form of energy with effectively negative CO_2 emissions, in that the system would pull CO_2 out of the air during the growing cycle of the plant and then permanently store it in some carbon repository, probably underground.

The IEA states that achieving both the best case for bio-fuels along with the best case for biomass power and heat in 2050 would require 8 to 11 billion dry metric tons of biomass. They note, "Studies suggest such supply could be sourced in a sustainable way from wastes, residues and purpose-grown energy crops." Bioenergy from food crops is not likely to be a major player.

If we do not get on the 2°C pathway, future available bio-mass supply for creating bioenergy is likely to be quite limited. In a globally warmed world, most available arable land will be used for growing food. Other than the United States, most countries do not have a lot of excess arable land, so even mod-erate Dust-Bowlification and sea-level rise and ocean acidifica-tion are likely to greatly limit their options for feeding their people. It seems likely the world will need yet another genera-tion of biofuels, one that uses exceedingly little arable land or fresh water, if either low-carbon biomass power or low-carbon biofuels (or both) are to become major contributors to avoiding dangerous climate change.

What other carbon-free forms of energy can contribute to cutting greenhouse gas emissions?

On the 2°C path, the world has to replace vast amounts of fos-sil fuels with carbon-free energy in the coming decades. So governments and the private sector will likely pursue almost every conceivable form of low-carbon energy. Technologies that are not already commercialized or close to being com-mercialized are unlikely to play a big role in replacing fossil fuels. There are a number of sources of low-carbon electricity in particular that merit attention.

The leading source of renewable electricity worldwide is hydropower, principally hydroelectric dams. Hydropower supplies more than 16% of the world's electricity, which is more than nuclear power. The 2012 International Energy Agency report "Technology Roadmap: Hydropower" notes that, "Since

2005, new capacity additions in hydropower have generated more electricity than all other renewables combined." The IEA report envisions a doubling of global hydropower capacity and electricity generation, most of which will come from big hydroelectric projects in developing countries (and some will come from repowering older dams with better turbines, a highly cost-effective strategy). Because global demand for electricity will likely rise substantially in the coming decades, even a doubling of hydropower's output will result in its share of global electricity only increasing slightly to 20%.

Geothermal is another important form of renewable energy. Geothermal comes in two types: large-scale power plants and smaller-scale systems, such as heat pumps. Large-scale plants use hot water or steam directly from under the ground to spin a turbine and produce electricity. Geothermal provides a large percentage of "total electricity demand in Iceland (25%), El Salvador (22%), Kenya and the Philippines (17% each), and Costa Rica (13%)." There are some 11,000 megawatts (11 GW) of capacity built around the world, which is a very small fraction of global electricity. However, just as the natural gas industry has seen a renaissance because of new drilling techniques that let drillers access unconventional gas cost-effectively, the geothermal sector may similarly see vast new resources open up. At this time, however, it is unclear whether large-scale geothermal power could be a major contributor (>5%) to a very low-carbon global energy system. The Intergovernmental Panel on Climate Change has estimated it could contribute 4%. Likewise, the IEA's 2011 "Technology Roadmap: Geothermal Heat and Power" report foresees that with concerted effort, geothermal electricity could reach 3.5% of global electricity generation in 2050.

Smaller-scale geothermal systems use the underground heat for direct heating of buildings and to assist various industrial processes. There are some 30 GW of such systems around the world, half of which use geothermal heat pumps that provide both heating and cooling. They generally run piping a few hundred feet below the Earth's surface. There, the temperature stays relatively constant throughout the year. Because the ground is warmer than the outside air in the winter, the geothermal system has to expend less energy than conventional systems to heat a building. Similarly, because the ground is colder than the outside air during the summer, the geothermal system has to expend less energy than conventional systems to cool a building. Therfore, geothermal heat pumps maintain very high efficiency all of the time in virtually any climate. Geothermal heat pumps have a high upfront cost but have seen steady growth in recent years because in many applications in many parts of the world, they are the most efficient form of heating and cooling available. The IEA projects that direct geothermal heating could provide almost 4% of predicted final energy used for heating in 2050. Also, because they are an electric heat pump, they can be paired with on-site renewable electricity, such as solar photovoltaics, and provide carbon-free heating and cooling. In a 2°C world, such a combination may prove popular.

There are many other low carbon forms of electricity, particularly ones that make use of water. These include extracting energy from the tides and from waves, as well as ocean thermal energy conversion, which uses the temperature difference between warm shallow water and cooler deep water to generate electricity. Currently, these are niche technologies whose future potential is unknown, but at this point, they do not seem significant relative to the scale of the energy transition required to stabilize at or near 2°C.

Finally, fusion remains a popular long-term hope for carbon-free power, as it has for a half-century now. Re-creating the sun's power source on Earth in a practical and cost-effective fashion has proved intractable even after decades of research. We currently do not know whether practical or affordable fusion is possible. The *New York Times* editorialized in 2012 about the high price of the current U.S. fusion program, "If the main goal is to achieve a power source that could replace fossil fuels, we suspect the money would be better spent on renewable sources of energy that are likely to be cheaper and quicker to put into wide use."[58] Put another way, harnessing the power of fusion, from 93 million miles away, is likely a better investment.

How can we reduce carbon dioxide emissions in the transportation sector?

Any climate policy effort must have a significant focus on transportation, a major source of human-caused carbon dioxide emissions. Globally, transportation accounts for one quarter of all energy-related carbon dioxide emissions and half of all oil consumed.[59]

Transportation may be the single most challenging sector for emissions reductions. "Transportation, driven by rapid growth in car use, has been the fastest growing source of CO_2 in the world," as the coauthor of an Institute for Transportation and Development Policy report explained in 2014. The U.S. Energy Information Administration similarly reports, "The transportation sector has dominated the growth in U.S. carbon dioxide emissions since 1990, accounting for 69% of the total increase in U.S. energy-related carbon dioxide emissions." The European Commission reports "Transport is the only major sector in the EU where greenhouse gas emissions are still rising."

The United States and the EU have been able to reduce the other dominant source of energy-related carbon dioxide emissions—power generation. In part, that is because there are so many carbon-free sources of electricity, such as solar, wind, nuclear power, and hydropower. In addition, many of those sources, particularly the high-tech renewable ones such as wind and solar, have come down sharply in price as production has increased. However, globally, 95% of all energy consumed by our cars, sport utility vehicles, vans, trucks, ships, trains, and airplanes is still petroleum-based—just as it has been for a long time. That is because the overwhelming

majority of our vehicles are still powered by internal combustion engines running on liquid fossil fuels—just as they
have been for a century. Also, the total number of vehicles
worldwide has grown inexorably for decades. Motor vehicles
alone have doubled in the last quarter century to more than
1 billion.

The kind of deep global reductions in carbon dioxide emissions needed to avoid warming the planet 2°C will require
multiple strategies in the transportation sector just as it has in
other sectors. Four of the most widely-used strategies for cutting transportation-related emissions are as follows:

1. Energy efficiency: improving the fuel economy of
   vehicles
2. Alternative fuels vehicles: cars that run on low-carbon
   fuels
3. Substitution: replacing personal vehicle travel with
   walking, biking, public transport, or even potentially the
   Internet.
4. Gasoline taxes: to encourage consumers to choose one or
   more of the above strategies

To date, efficiency has been the most widely used and most
cost-effective technology strategy. It is the simplest approach
for reducing vehicle emissions without requiring a new fueling infrastructure. Another key advantage is that fuel economy
improvements can pay for themselves by reducing fueling
costs. The payback for efficiency is faster when the consumer
price of gasoline is higher.

The energy crises of the 1970s dramatically increased
the price of gasoline and motivated many of the developed
countries to adopt policies to promote efficiency. The United
States embraced fuel economy standards, whereas European
countries embraced fuel taxes. Eventually, many countries
in the world, including China, adopted efficiency standards.

Automakers have pursued many strategies to improve vehicle fuel efficiency. These include making the vehicles lighter, improving aerodynamics (so there is less fuel wasted overcoming drag), switching to diesel engines (which are considerably more efficient than gasoline-powered internal combustion engines), and using other more advanced technologies. One of the most important of these advanced technology strategies is hybrid vehicles, which combine an electric motor with the traditional gasoline (or diesel) engine. This combination provides opportunities for significant fuel savings. For instance, a major energy saver is that the car can now use "regenerative braking," whereby the motor captures some of the energy that is lost to frictional heat in most vehicles when they slow down. Hybrids capture approximately half of the energy lost in braking.

Another road for avoiding or reducing transportation-related carbon pollution is substitution. Substitution includes things like mass transit, bus systems, ridesharing, bike paths, and the like. Perhaps the biggest opportunity for substitution comes from the Internet. There are already indications that the Internet is beginning to affect the amount of total annual driving in industrialized countries like the United States, because more people telecommute, teleconference, and tele-shop on websites like Amazon. There is evidence that younger people are buying fewer cars and driving less as texting, Facebook, and social media become their preferred means of communications and socializing.

Although more efficient vehicles and reductions in vehicle miles traveled are important, the world needs to cut transportation emissions sharply even as the number of vehicles on the road rises. Therefore, we will need to replace gasoline with a zero-carbon fuel, which means we will need to replace conventional vehicles with alternative fuel vehicles (AFVs), such as electric vehicles or hydrogen fuel-cell vehicles. Let us look at the challenges that AFVs face.

What challenges have limited the marketplace success
of alternative fuels and alternative fuel vehicles to date?

Alternative fuel vehicles (AFVs) are vehicles that run on a fuel other than petroleum-based fuels such as gasoline or diesel. Examples of alternative fuels are electricity, natural gas, and hydrogen. Alternative fuel vehicles face many challenges—even after decades of efforts to advance AFVs, some 95% of the fuel used by U.S. vehicles is still petroleum-based. In particular, compared with conventional vehicles running on conventional fuels, AFVs typically suffer from several marketplace disadvantages, such as higher cost and difficulties with vehicle refueling. Hence, they inevitably require government incentives or mandates to succeed. Second, they typically do not provide cost-effective solutions to major energy and environmental problems, which undermines the policy case for having the government intervene in the marketplace to support them.[60]

The limitations of AFVs as climate solutions have been understood for a long time. Consider the 2003 analysis by the U.S. Department of Transportation, "Fuel Options for Reducing Greenhouse Gas Emissions from Motor Vehicles." The report assessed the potential for gasoline substitutes to reduce GHG emissions over the next 25 years. It concluded that "the reduction in GHG emissions from most gasoline substitutes would be modest" and that "promoting alternative fuels would be a costly strategy for reducing emissions." It has always been hard to beat fuel economy as the most cost-effective strategy for cutting emissions.

The U.S. government, the EU, and others (such as California and Canada) have tried to promote AFVs for a long time. America's 1992 Energy Policy Act established the goal of having alternative fuels replace at least 10% of U.S. petroleum fuels in 2000 and at least 30% of U.S. petroleum fuels in 2010. The United States was at 1% alternative fuels in 2000 and did not even hit the 2000 target in 2010. In addition, the primary alternative fuel, as discussed above, is corn-based ethanol blended into gasoline, and such ethanol offers little GHG benefit. Corn ethanol has succeeded primarily to the extent that relatively low levels of it (up to 10%) can be blended directly with gasoline and used in most modern cars.

A significant literature has emerged explaining this failure. Besides the question of whether AFVs deliver cost-effective emissions reductions, there have historically been six major barriers to their success:

1. **High first cost for vehicle:** Can the AFV be built at an affordable price for consumers and still be profitable for the manufacturer?
2. **On-board fuel storage issues (i.e., limited range):** Can enough alternative fuel be stored onboard to give the car the kind of range consumers expect, without compromising passenger or cargo space? Can the AFV be refueled fast enough to satisfy consumer expectations?
3. **Safety and liability concerns:** Is the alternative fuel safe, something typical users can easily handle without special training?
4. **High fueling cost (compared to gasoline):** Is the alternative fuel's cost (per mile) similar to (or cheaper than) gasoline? If not, how much more expensive is it to use?
5. **Limited fuel stations (the chicken and egg problem):** On the one hand, who will build and buy the AFVs in large quantity if a broad fueling infrastructure is not in place to service them? On the other hand, who will build that fueling infrastructure, taking the risk of a massive stranded investment, before a large quantity of AFVs are built and bought, that is, before these particular AFVs have been proven to be winners in the marketplace?
6. **Improvements in the competition:** If the AFV still needs years of improvement to be a viable car, are the competitors, including fuel-efficient gasoline cars, likely to improve as much or more during this time? In short, is it likely that competitors will still have superior vehicles in 2020 or 2030?

All AFVs thus far have suffered from several of these barriers. Any one of them can be a showstopper for an AFV or an alternative fuel, even where other clear benefits are delivered. The most intractable barrier remains the chicken and egg problem: who will build and buy the AFVs if a fueling infrastructure is not in place to serve them and who will build the fueling infrastructure before the AFVs are successful in the market. Consider the fact that there are millions of flexible fuel vehicles already on the road that are capable of running on E85 (85% ethanol, 15% gasoline), 100% gasoline, or just about any blend, for about the same price as gasoline-powered vehicles; however, the vast majority of them run almost entirely on gasoline.

In the case of cars powered by natural gas, the environmental benefits were oversold, as were the early cost estimates for both the vehicles and the refueling stations: "Early promoters often believe that 'prices just have to drop' and cited what turned out to be unachievable price levels." One study concluded, "Exaggerated claims have damaged the credibility of alternate transportation fuels, and have retarded acceptance, especially by large commercial purchasers." Electric vehicles deliver zero tailpipe emissions, and they are the only alternative vehicle to have substantially lower per mile costs than gasoline cars, but range, refueling, and first-cost issues have, until recently, limited their success. Hydrogen fuel cell vehicles (FCVs) also deliver zero tailpipe emissions, but they suffer from all of the barriers described above. Fuel-cell vehicles remain the most difficult and expensive kind of AFV currently under consideration.

One problem that slowed the penetration of all AFVs was increasing competition from improved gasoline-power vehicles. For instance, decades ago, states such as California and the United States as a whole were not focused on global warming. They simply wanted to clean up the air, especially in the most polluted cities, for the sake of public health. So governments introduced tighter tailpipe emissions standards for urban air pollutants such as the oxides of nitrogen (NOx) and volatile organic compounds, both of which create smog, especially on hot summer days. However, when those standards were being developed, few suspected that requiring cars to emit no more than 0.02 grams of NOx per mile could be achieved by improved internal combustion engine vehicles running on improved (reformulated) gasoline.

However, in the 2000s, automakers introduced a new generation of vehicle, the hybrid partial-zero emission vehicle (PZEV), such as the Toyota Prius and Ford Escape hybrid. Those vehicles substantially raised the bar for any potential AFV. The hybrid PZEVs have no chicken and egg problem (since they run on gasoline and can be fueled everywhere), a substantially lower annual fuel bill, greater range, a 30%–50% reduction in GHG emissions, and a 90% reduction in tailpipe emissions. Toyota in particular addressed all emissions problems (and the oil consumption problem) with its remarkable 2004 Prius, arguably the world's first truly practical and affordable (i.e., mass market) green car. The Prius was not merely almost twice as fuel-efficient as cars in its class, it was also a partial-zero emission vehicle with substantially lower emissions of regulated air pollutants compared with a typical new car at the time.

Hybrids do cost a little more, but that increase has generally been offset by the government incentive and the large reduction in gasoline costs, even ignoring the performance benefits. With most alternative fuel vehicles, however, the environmental benefits, if any, typically come at the expense of a higher first cost for the vehicle, a much higher annual fuel bill, a reduced range, and other undesirable attributes from the consumer's perspective. Therefore, from a green consumer perspective, if you wanted a car that slashed oil consumption, slashed carbon pollution, slashed tailpipe emissions, was affordable, and could be fueled everywhere, and a car that also required half the trips to the gas station and had a very large range on one tank of gasoline, then you could buy a Prius. If not, you could also buy some of the other PZEV hybrids, or you could buy a conventional PZEV that was fuel-efficient, as many people did.

Those advances in gasoline-fueled cars definitely undercut the consumer value proposition and government motivation for purchasing AFVs, including fuel cell and electric vehicles. How much extra cost plus inconvenience would the average car buyer be willing to accept just to reduce tail pipe emissions from the already very low 0.02 grams of NOx per mile down to 0.00 (and not to reduce evaporative emissions at all)? The answer is, unsurprisingly, not much. However, over the past decade or so, whereas concern over new car tailpipe emissions dropped somewhat, concern over carbon dioxide emissions has grown, not just in the U.S. but globally, where most new cars are sold. The growing interest in climate solutions meant renewed interest in a true zero-emission vehicle: one that did not just have zero tailpipe emissions, but one that ran on a fuel that itself was virtually emission-free and carbon-free at the source—since both electricity and hydrogen can be produced in a variety of ways, some very clean and others quite dirty.

At the same time, manufacturers and entrepreneurs kept working to improve electric vehicle and fuel-cell technology. The Bush Administration launched a hydrogen car initiative that dramatically boosted spending on hydrogen FCVs. In addition, massive amounts of money were poured into improving batteries and related components, not just by governments, car makers, and clean energy venture capitalists, but also by portable device and phone manufacturers who wanted to improve performance of their much smaller devices while cutting costs. In recent years, both EVs and FCVs have achieved somewhat of a comeback, especially EVs. Let us look at each in turn.

What role can electric vehicles play?

Electric vehicles are vehicles with an electric motor and a large enough battery to run the vehicle a substantial distance before recharging. The earliest cars running on an electric drive date back to the 1830s. As late as 1900, one third of the cars on U.S. roads were electric. However, the discovery of vast amounts of oil in the country, coupled with better roads, better engine technology, and mass production, made it very hard for EVs to compete with internal combustion engine vehicles in most applications.[61]

With growing concerns about both oil dependence and pollution in the 1970s and 1980s, there was renewed interest in EVs. EVs have long been one of the only Alternative Fuel Vehicles with a per-mile fueling cost far below that of gasoline. That is because electric motors are highly efficient, and there are many low-cost sources of electric power. Of particular interest in the climate arena is the fact that electricity from renewable sources (or nuclear power) is the only zero-carbon alternative that is substantially cheaper than gasoline—and this was true even before the huge recent declines in the cost of renewable power, including a 99% plunge for solar power in the past quarter-century.

To make use of low-cost carbon-free power, however, you still need a practical electric car. Steady improvements in battery technology have turned out to be a game-changer over time, especially when combined with design innovations. The first big change came with the success of hybrid-electric cars such as the Toyota Prius. Hybrids demonstrated that a car integrating gas and electric drives can be practical, affordable, and desirable. Even so, hybrid vehicle batteries are still small, so they do not take up that much space or add that much weight and cost. In addition, they charge either using electricity regenerated during braking or electricity generated from the gasoline engine, the latter of which does not solve the problem of pollution.

The success of gasoline hybrids in turn led to an entirely different kind of EV: the plug-in hybrid electric vehicle. Improvements in gasoline hybrids, including a bigger and better battery, allowed them to be plugged into the electric grid and run in an all-electric mode for a modest range before recharging. Because most vehicle use is for relatively short trips, such as commuting or shopping, followed by an extended period of time during which the vehicle is not being driven and could be charged, even a relatively limited all-electric range of 20 or 40 miles would allow these vehicles to replace a substantial portion of gasoline consumption and tailpipe emissions.

However, because plug-in hybrids have a gasoline engine, and thus are dual-fuel vehicles, they avoid two of the biggest problems of pure EVs. First, they are not limited in range by the total amount of battery charge with which they start. If their initial charge runs down, the vehicle can run purely on gasoline and whatever charge is possible with regenerative braking. Second, instead of taking many hours to charge as EVs have traditionally done, the plug-in can refuel within minutes just like a gasoline car, if for some reason the owner does not have the time between trips to charge up or if they cannot find a local charging session.

From 2009 through 2014, more than 260,000 plug-in hybrids were sold globally. The world's top-selling plug-in hybrid is the General Motors Chevrolet Volt (and similar cars sold under a different brand). Total sales exceed 88,000. The Toyota Prius plug-in is the number two seller, with more than 65,000 sold.

Another game changer in the recent history of electric vehicles has been the emergence of Tesla Motors. The company was founded in 2006 as a Silicon Valley startup by Elon Musk to build a high-end electric sports car with a single-charge range of over 200 miles. By 2014, Teslas had become California's "largest auto industry employer," according to the U.S. Department of Energy. Its market capitalization (the total value of all its stock) was more than half that of GM's, despite having a small fraction of GM's sales or revenues.

The DOE explains in its history of EVs, "Tesla's announcement and subsequent success spurred many big automakers to accelerate work on their own electric vehicles." By the end of 2010, Chevy had released the Volt plug-in hybrid in the United

States. At the same time, Nissan released the LEAF, which has become the top selling all-electric car on Earth, with total sales of over 165,000 as of early 2015. Tesla's Model S has sales of more than 50,000.

The Nissan LEAF has an overall fuel economy equivalent to 115 miles per gallon. Its range is 120 miles (200 km). Most of the growth in car use in the coming decades will occur in countries where people do not necessarily drive long distances on a regular basis and do not require large vehicles. The LEAF is used for city driving and short trips, and, in many cases, it is driven more than the primary vehicle, because the majority of trips most people take are short.

"Before 2010, there was effectively no demand for electric vehicles," the U.S. DOE notes. By the end of 2014, more than 700,000 total plug-in vehicles had been sold worldwide (plug-in hybrids and pure battery electrics), up from approximately 400,000 at the end of 2013. Some 320,000 EVs were registered in 2014. As of 2015, more than 30 models of commercial cars and vans powered purely by electricity are now available for purchase, primarily in Europe, the United States, Japan, and China.

A key reason for the sharp increase in interest in EVs is the sharp drop in the cost of their key component, batteries. The energy stored in a battery is measured by kilowatt-hour (kWh), which is equal to 1 kW of power delivered for 1 hour. The more kWh stored, the farther the car can go on one charge. A key metric for battery economics is the cost per kWh. The lower cost, the cheaper it is to build an electric car with a significant range. In a major 2013 analysis, "Global EV Outlook: Understanding the Electric Vehicle Landscape to 2020," the International Energy Agency estimated that EVs would achieve cost parity with internal combustion engine vehicles when battery costs hit $300 per kWh of storage capacity. The analysis projected that would happen by 2020.

Yet, a March 2015 study in *Nature Climate Change*, "Rapidly falling costs of battery packs for electric vehicles" determined that "industry-wide cost estimates declined by approximately 14% annually between 2007 and 2014, from above US$1,000 per kWh to around US$410 per kWh." The study went even further to note, "the cost of battery packs used by market-leading BEV [Battery Electric Vehicle] manufacturers are even lower, at US$300 per kWh."

In short, the best manufacturers have already reached the battery price needed for cost parity with conventional cars. A 2014 report from UBS, a leading investment bank, found "the 3-year total cost of ownership of a Tesla S model is similar to that of a comparable petrol combustion engine car such as an Audi A7," in places such as Germany. Of course, parity was based on a 2013 and 2014 price for oil that was considerably higher than the price for oil by 2015. The 2015 battery study found that battery prices would need to drop under $250 per kWh for EVs to become competitive. Furthermore, it concluded:

> If costs reach as low as $150 per kilowatt hour this means that electric vehicles will probably move beyond niche applications and begin to penetrate the market more widely, leading to a potential paradigm shift in vehicle technology.

Can EV batteries hit that price point? The 2015 study projects that costs will fall to some $230 per kWh in the 2017 to 2018 timeframe. Tesla Motors and Panasonic have started building a massive $5 billion plant that can produce half a million battery packs each year (plus extra batteries for stationary applications). It is expected to be completed in 2017. Tesla and Panasonic estimate this 6500-employee "Gigafactory" will lead to a 30% reduction in cost, which the 2015 study said is "a trajectory close to the trends projected in this paper." It may well be that $150 per kWh can be achieved by 2020 without a major battery breakthrough but simply with continuing improvements in manufacturing, economies of scale, and general learning by industry.

What about the range and refueling problem? Several car companies, including GM, Ford, Nissan, and Volkswagen, have announced that they will sell an EV with a 200-mile range. In addition, although charging at typical EV stations can still take hours, the superfast charging stations some companies are now building can do a substantial charge in 25 minutes or less. As of 2015, the fastest chargers can charge a battery up to 80% of its capacity in 20 minutes. Therefore, fast charging is reaching the speed needed for the kinds of rest area stops many people make on trips. The EV has not solved the range issue entirely, but it is in the process of being addressed enough for many applications, including the city cars that are popular in many countries. In addition, there is every reason to believe that battery and charging technology will continue their steady improvement, further opening up potential markets for EVs.

What are hydrogen fuel cells and hydrogen fuel-cell vehicles?

A fuel cell is a small, modular electrochemical device, similar to a battery in many respects, but which can be continuously fueled. You can think of a fuel cell as a black box that produces water and electricity and heat when you feed it hydrogen and oxygen. Fuel cells can be used in both stationary and mobile applications as long as there is a steady supply of hydrogen. Fuel cell vehicles are like electric cars in many respects because fuel cells put out electricity that runs an electric motor. The main difference is that FCVs require a source of hydrogen to operate, and EVs need a source of electricity.[62]

Fuel cells were first invented in the 1830s. For over a century, they had virtually no marketplace success. Nonetheless, they have proven to be very useful for space missions such as Apollo because outer space is not a good place for combustion. However, those fuel cells were not cost-effective or practical for use as commercial products. Moreover, it has only been in very recent years that fuel cells have been commercialized for niche applications. The most promising fuel cell for transportation uses is the proton exchange membrane (PEM), which was first developed in the early 1960s by General Electric for the Gemini space program. These PEM fuel cells operate at a low temperature, which means they do not take much time to warm up. Many of the fuel cells that have been developed and are being used today are high temperature fuel cells for stationary applications, which are not practical for cars.

One of the first briefings I received when I came to the Department of Energy in mid-1993 was on some break-throughs at Los Alamos National Laboratory and elsewhere that brought down the cost of PEM fuel cells. Proton exchange membrane fuel cells were still nowhere near affordable, but a time when they might be had become imaginable. So, in the 1990s, the Department sharply increased funding for hydrogen and for these PEM fuel cells. At the same time, we were spending a lot of money pursuing hybrid vehicles, advanced batteries, and many alternative fuel vehicles. Hydrogen fuel-cell cars had long been considered the Holy Grail of AFVs because they have zero tail-pipe emissions, like EVs, but can be continuously refueled quickly if there is a source of hydrogen nearby. Thus, they could potentially eliminate the range and refueling problem that had plagued EVs.

In 2003, President George W. Bush announced in the State of the Union that he was calling on the nation's scientists and engineers to develop a hydrogen fuel cell car so that, for a child born in that year, their first car would run on hydrogen and be pollution free. That set off a massive increase of spending by the federal government and investments by private companies in hydrogen and fuel cells.

When I was at the Department, one of the reasons we were excited about hydrogen is that we were pursuing technology to convert gasoline to hydrogen onboard the car, which is called "onboard reforming." However, it became clear in the 1990s that onboard reforming was not practical for a car, so the government and most private companies essentially gave up on that idea. Yet, I am not sure we realized the full implications of doing so at the time. It meant that large-scale introduction of hydrogen fuel cell cars required someone to build thousands of expensive hydrogen fueling stations around the country to give people confidence they could refuel when needed. It also meant that we would need to store large amounts of hydrogen onboard the car. In addition, all of this meant that building a practical and affordable hydrogen car was going to be much, much more difficult than anybody realized.

After the President's announcement, I began researching a hydrogen primer. As I read the literature and talked to the experts and did my own analysis, it became increasingly clear that hydrogen cars were a very difficult proposition. My book, "The Hype About Hydrogen" came out in 2004. At the same time, the National Academy of Sciences came out with a study that was equally sobering, and so did the American Physical Society. My conclusion in 2004 was that "hydrogen vehicles are unlikely to achieve even a 5% market penetration by 2030." So hydrogen fuel cell cars were not going to be major contributor to addressing climate change. A 2013 study by independent research and advisory firm Lux Research came to a similar conclusion. Their study "The Great Compression: The Future of the Hydrogen Economy" concluded that despite billions in research and development spent in the past decade, "The dream of a hydrogen economy envisioned for decades by politicians, economists, and environmentalists is no nearer, with hydrogen fuel cells turning [into] a modest $3 billion market of about 5.9 GW in 2030."

For the purposes of this book, I am not discussing stationary fuel cells as a climate solution, because there is little reason to believe they will be a major contributor. Developing a successful commercial stationary fuel cell has proven to be more difficult than anyone anticipated. There are very few profitable commercial ventures making such fuel cells. The 2013 Lux study concluded that PEM fuel cells in niche applications (telecom and backup power) will have a $1 billion market in 2030, but that "fuel cells of all types for residential, commercial and

utility generation will not prove cost-effective." That last conclusion may be too pessimistic, but stationary fuel cells, especially the more efficient high-temperature ones best suited for
buildings and power generation, have another problem. Now
and for the foreseeable future, they are almost entirely fueled
by natural gas to keep costs as low as possible. However,
in most applications, such fuel cells offer little or no carbon
savings over the competition. In fact, in many applications,
stationary fuel cells running on natural gas generate more
emissions than existing alternatives.

Lux projects that FCVs and other mobile applications will
be a $2 billion a year market in 2030. Unfortunately, hydrogen
FCVs will not be a major contributor to solving the problem
of manmade climate change until (1) the market is 100 times
larger than Lux projects and (2) affordable carbon-free hydrogen is ubiquitous, which does not appear likely to happen fast
enough to matter to in the effort to sharply cut greenhouse
gas emission in the next few decades. Lux itself concluded that
"hydrogen demand from fuel cells will total 140 million kg in
2030, a meager 0.56% of global hydrogen demand."

Given that a few major car companies, including Toyota, are
in the process of introducing FCVs to the consumer market, let
us look at the challenges they face.

What are the challenges facing hydrogen fuel-cell vehicles as a climate solution?

Fuel-cell vehicles face all six problems shared by one or more
alternative fuel vehicles discussed above: high first cost,
onboard fuel storage issues, safety and liability concerns, high
fueling cost (compared with gasoline), limited fuel stations,
and improvements in the competition.[63]

First, fuel cell cars are very expensive. There has been
steady but slow success in lowering their cost over time, so
this remains a major stumbling block. Ford Motor Company
has been pursuing FCVs for more than a decade. By 2009 their

30 Ford Focus FCVs had, combined, over a million miles on them. After all this work, Ford's *Sustainability Report 2013/2014* explained:

> Even with the advances we have made in hydrogen technology over the past 10 years, we still have challenges to overcome before hydrogen FCVs can compete in the market with current vehicle technology. The cost and durability of the fuel cell system are the most significant challenges. For example, extensive DOE analysis has not yet revealed an automotive fuel cell technology that meets the DOE's targets for real-world commercialization, or that maintains proper performance throughout the targeted lifetime while staying within the targeted cost.

Therefore, after a decade of research, development, and demonstration, which itself built on a decade of R&D at DOE and elsewhere, we still do not have a practical, affordable fuel-cell technology.

Other companies face similar problems, but they have decided to start marketing vehicles anyway. Toyota's new FCV, the Mirai, will sell for some $70,000 before incentives. However, Pat Cox, former President of the European Parliament, said in a November 2014 talk on hydrogen cars that Toyota will probably lose between 50,000 and 100,000 euros ($66,000 and $133,000) on each car. It is always difficult to get reliable numbers on the actual cost of a new technology, but in February 2015, Hyundai said that it would slash the price of its Tucson SUV 43% to compete with Toyota. The vehicle originally cost some $144,000 and the price cut is over $60,000. The actual cost to produce Toyota's new FCV, the Mirai, and Hyundai's FCV, the Tucson, could be well over $100,000.

Second, fuel-cell cars have still not solved their onboard storage issue, which is key for them to be a practical vehicle. At room temperature and pressure, hydrogen takes up some 3,000 times more space than gasoline containing an equivalent amount of energy. A decade ago, the U.S. Department of Energy's 2003 *Fuel Cell Report to Congress* explained that, "Hydrogen storage systems need to enable a vehicle to travel 300 to 400 miles and fit in an envelope that does not compromise either passenger space or storage space. Current energy storage technologies are insufficient to gain market acceptance because they do not meet these criteria."

The "current energy storage technologies" at the time were liquefied hydrogen and compressed hydrogen gas. Liquid hydrogen is widely used today for storing and transporting hydrogen. Indeed, for storage and fueling, liquids enjoy considerable advantages over gases: they have high-energy density, are easier to transport, and are typically easier to handle. That is why the overwhelming majority of vehicles today are powered by liquids such as gasoline and diesel.

However, liquid hydrogen is anything but typical. It becomes a liquid only at minus 423 degrees Fahrenheit, just a few degrees above absolute zero. That in turn requires it to be stored in a super-insulated cryogenic tank. Liquid hydrogen is very unlikely to be used in vehicles because of the cost and logistical problems in handling it and also because liquefaction is so energy-intensive. The process of liquefying hydrogen for storage requires some one third of the energy of the hydrogen!

Because the alternatives were so poor, nearly all prototype hydrogen vehicles used compressed hydrogen storage a decade ago. Hydrogen is compressed up to pressures of 10,000 pounds per square inch (psi) in a multistage process that requires energy input equal to 10% to 15% of the hydrogen's usable energy content; to compare, atmospheric pressure is approximately 15 psi. Working at such high pressures makes the entire storage system, including the fuel pumps, very complex. It requires materials and components that are complicated and costly. In addition, even a 10,000-psi tank takes up seven to eight times the volume of an equivalent-energy gasoline tank or perhaps four times the volume for a comparable range (because the FCV will be more fuel efficient than current cars).

The 2004 National Academies of Science study on "The Hydrogen Economy" concluded that both liquid and compressed storage have "little promise of long-term practicality for light-duty vehicles" and recommended that DOE halt research in both areas. Practical hydrogen storage requires a major technology breakthrough, most likely in solid-state hydrogen storage. Yet, a decade later, nearly all hydrogen demonstration vehicles as well as those planned for market introduction still use compressed hydrogen storage. As Ford explained in its 2013/2014 report:

> On-board hydrogen storage is another critical challenge to the commercial viability of hydrogen FCVs. Current demonstration vehicles use compressed gaseous hydrogen storage. However, the high-pressure tanks required for this storage use expensive materials for reinforcement such as carbon fiber. In addition, the current tanks are large and difficult to package in a vehicle without unacceptable losses in passenger or cargo space.

Ultimately, Ford concludes that because of the "still significant challenges related to the cost and availability of hydrogen fuel and onboard hydrogen storage technology," we need "further scientific breakthroughs and continued engineering refinements" in order to "make fuel cell vehicle technology commercially viable."

The safety issues for hydrogen are unique because the characteristics of hydrogen are so unusual. On the one hand, the gas does have some advantages compared with liquid fuels such as gasoline. When a gasoline tank leaks or bursts, the gasoline can pool, creating a risk that any spark would start a fire, or it can splatter, which risks spreading an existing fire. By contrast, hydrogen escapes quickly into the air as a very diffuse gas. Also, hydrogen gas is nontoxic. However, hydrogen has its own special safety issues. The DOE in its discussion of FCV safety, notes "Hydrogen is also odorless, colorless, and tasteless making it undetectable by human senses. For these reasons, hydrogen systems are designed with ventilation and leak detection." The addition of common odorants such as sulfur is impractical, in part because they poison fuel cells.

Hydrogen is highly flammable, with an ignition energy that is 10 times smaller than that of natural gas or gasoline. It can be ignited by cell phones or by electrical storms located miles away. Hence, leaks pose a significant fire hazard, particularly because they are hard to detect. Hydrogen burns nearly invisibly, and people have unwittingly stepped into hydrogen flames. Hydrogen can cause many metals, including the carbon steel widely used in gas pipelines, to become brittle. In addition, any high-pressure storage tank presents a risk of rupture. More than one fifth of hydrogen accidents are caused by undetected leaks.

The DOE states that "Hydrogen can be used as safely as other common fuels we use today when guidelines are observed and users understand its behavior." In general, the hydrogen industry makes extensive use of special training, special clothing, and electronic flame gas detectors. In industry, hydrogen is subject to strict and cumbersome codes and standards, especially when used in an enclosed space such as a laboratory, or a garage, where a leak might create a growing gas bubble. Moreover, accidents still occur. What would happen if there were widespread use of hydrogen cars by people with unventilated garages? The former group leader for energy storage programs at Ford has argued "it is difficult to imagine how hydrogen risks can be managed acceptably by the general public when wide-scale deployment of the safety precautions would be costly and public compliance impossible to ensure." It seems likely that major innovations in safety will be required before a hydrogen economy is practical.

Another key issue is the infrastructure problem. Hydrogen fueling stations are very expensive, generally $1 million to $2 million, because safely pressurizing hydrogen to 10,000 pounds per square inch pressure is not easy. In addition, many stations receive (and store) hydrogen as a super-cooled liquid, which requires great care. The cost is a key reason why there are so few stations nationwide. Stations that produce carbon-free hydrogen can be even more expensive, which is why there are far fewer of them.

The high cost of hydrogen fueling stations raises several questions. Who will spend the tens, if not hundreds, of billions of dollars on a wholly new nationwide infrastructure to provide ready access to carbon-free hydrogen for consumers with FCVs until millions of hydrogen vehicles are on the road? On the other hand, who will make and market such expensive vehicles until the infrastructure is in place to fuel them? Moreover, will car companies and fuel providers be willing to take this chance before knowing whether the public will embrace these cars?

Thus "the risk of stranded investment is significant, since much of an initial compressed hydrogen station infrastructure could not be converted later," if a better means of onboard storage than very high pressure canisters is developed or if someone comes up with a better fuel for FCVs, as a 2001 study explained. The risk of stranded investment is also great for fueling stations that are powered by purified natural gas (methane or CH_4) because it is relatively easy and cheap to extract the hydrogen. The key problem is that this process releases carbon dioxide.

Ford's "Sustainability Report 2013/2014" notes that "Currently, the most state-of-the-art procedure is a distributed natural gas steam-reforming process. However, when FCVs are run on hydrogen reformed from natural gas using this process, they do not provide significant environmental benefits on a well-to-wheels basis (due to GHG emissions from the natural gas reformation process)." Such fueling stations would likely be of no value once the ultimate transition to a pure hydrogen economy was made, because that would almost certainly rely on centralized production and not make use of small natural gas reformers. Moreover, it is possible that the entire investment would be stranded in the scenario in which hydrogen cars simply never achieve the combination of popularity, cost, and performance to triumph in the marketplace.

Rather than have stations generate hydrogen, some have proposed that the hydrogen be delivered. Centralized production of hydrogen near abundant sources of carbon-free power (such as wind turbines) is the ultimate goal. The problem there is that tanker trucks carrying liquefied hydrogen are widely used to deliver hydrogen today, yet they have little value in a world trying to operating efficiently and cut carbon pollution at the lowest cost. That is because of liquefaction's high energy cost (a third of the energy contained in the hydrogen). In addition, almost all automakers today are using high-pressure storage, which means that a fueling station would have to use an energy-intensive pressurization system. In that case, storage and transport alone would require some 50% of the energy in the hydrogen delivered, which would wipe out most if not all of the energy and CO_2 benefits from hydrogen. As for simply delivering the hydrogen in trailers carrying compressed hydrogen canisters, a study by ABB concluded that the delivery energy approaches 40% of the usable energy in the hydrogen delivered if the travel distance is 300 miles.

Major breakthroughs in hydrogen production and delivery will be required to have affordable carbon-free hydrogen efficiently delivered to the car's onboard storage system. Ford's "Sustainability Report 2013/2014" explains:

There are also still significant challenges related to the cost and availability of hydrogen fuel and onboard hydrogen storage technology. To overcome these challenges and make fuel cell vehicle technology commercially viable, we believe further scientific breakthroughs and continued engineering refinements are required.

What role can energy storage play in the transition off of fossil fuels?

A challenge for some forms of renewable power is that they are intermittent (or variable), which is to say they operate only when the sun shines or the wind blows. This is not true of hydropower or geothermal or concentrated solar thermal electric. However, it is true of the two fastest-growing forms of renewable power—solar photovoltaics and wind. Over time, as the contribution of those two power sources become bigger and bigger in a carbon-constrained world, the electric grid will need strategies for how to handle periods of time when demand is high but the wind is not blowing and the sun is not shining.

There are two primary ways the variability problem is being addressed. First, half or more of this problem is really a "predictability problem." If we could predict with high accuracy wind availability and solar availability 24 to 36 hours in advance, then electricity operators have many strategies available to them. For instance, operators could plan to bring online a backup plant that otherwise needs several hours to warm up. Or they could use demand response, which is paying commercial, industrial, and even residential customers to reduce electricity demand for a short time given a certain amount of advance warning. In fact, such prediction capability is already being developed. As a 2014 article titled "Smart Wind and Solar Power" in the magazine *Technology Review* put it, "Big data and artificial intelligence are producing ultra-accurate forecasts that will make it feasible to integrate much more renewable energy into the grid."

A second way to deal with the variability of wind and solar photovoltaics is to integrate electricity storage into the grid. That way, excess electricity when it is windy or sunny can be stored for when it is not. The biggest source of electricity storage on the grid today is "pumped storage" at hydroelectric plants. In such plants, water can be pumped from a reservoir at a lower level to one at a higher level when there is excess electricity or when electricity can be generated at a low cost. Then, during a period of high electricity demand, which is typically a period of high electricity price, water in the upper reservoir is allowed to run through the hydroelectric plant's turbines to produce electricity for immediate sale. In the International Energy Agency's *Technology Roadmap: Hydropower*, "Pumped storage hydropower capacities would be multiplied by a factor of 3 to 5," by 2050.

The round-trip efficiency for pumped storage—the fraction of the original energy retained after the water is pumped up and comes back down—is 70% to 85%. In other words, 15% to 30% of the original energy is lost, which is quite good as storage systems go. Consider if you wanted to use hydrogen as the way to store power, using electrolyzers to convert the electricity to hydrogen, then storing hydrogen onsite until the electricity is needed, and finally running the hydrogen through a fuel cell to generate electricity again. Losses would likely exceed 50%, perhaps by a lot. That is a great deal of premium low-carbon electricity to lose, which suggests that fuel cells will only be used for storage in niche applications for quite some time. On the other hand, the round-trip efficiency of storing electricity in batteries can be better than the round-trip efficiency of pumped storage. The problem has been that, until recently, batteries have been too expensive for them to be used on a wide scale in most storage applications. As discussed earlier, however, battery prices are coming down sharply, as huge investments are being made in various types of battery technologies by electric car companies and others, including utilities. Also, if electric vehicles continue their rapid growth trend and become widespread, then it may be possible to access their batteries during the more than 90% of the time they are parked. That would potentially allow electric cars to provide storage or other valuable grid services. More importantly, the investment bank UBS report (discussed earlier) projects that "the payback time for unsubsidised investment in EVs plus rooftop solar plus battery storage will be as low as 6-8 years by 2020," so this transition may start sooner than expected.

Finally, as the IEA and others have noted, concentrated solar thermal electric power plants can build in low-cost storage (of a heated fluid) with very low round-trip losses. There is a great potential for this type of solar power plant to become a major feature of the electricity grid, especially after the 2020s, if the world decides to make the investments and policy changes needed to stabilize carbon dioxide at levels that minimize danger to humanity. Whenever the world gets serious about replacing fossil fuels with low-carbon energy, it seems likely that a combination of the technologies and strategies discussed above will be able to incorporate very large amounts of renewable electricity into the electric grid cost-effectively.

What can the agricultural and livestock sector do to minimize climate change?

The agriculture and livestock sector is a major contributor to greenhouse gas emissions. There are three basic ways that the sector can reduce emissions. First, it can cut its direct emissions of GHGs, including carbon dioxide released from the combustion of fossil fuels. Second, it can alter its practices so as to keep more carbon in the soil. Third, food providers can change what food they produce, because the production of some crops and livestock generate considerably more GHGs than others.[64]

First, like most other sectors of the economy, the agricultural sector can cut CO_2 emissions by using more efficient pieces of equipment and by using cleaner energy. Farmers have some options most others do not. For instance, because wind turbines are so tall, farmers have been able to put a large number of them on their land while still being able to farm underneath them. As a result, many farmers have been able to augment their earnings by harvesting the wind. In addition, farmers that raise livestock often have considerable animal waste that emits methane, which can also be harvested and used onsite for power and/or heat generation.

Second, some forms of agricultural land management practices store and preserve more carbon in the soil than others. Modifying tillage practices has been shown in some instances to increase soil carbon storage, but more research needs to be done to identify the optimal strategies and exactly how much carbon could be stored. Similarly, biochar, which is animal and plant matter that has been transformed into charcoal to store carbon in the soil is another option that may be able to remove carbon dioxide from the atmosphere. A 2012 report reviewed the literature on the subject, including 212 peer-reviewed studies. The authors point out that biochar would be an effective carbon reduction strategy only if it were stable in the soil for a long time. Otherwise, it would decompose and release its stored carbon back into the air. Their review "found that the data do not yet exist to accurately estimate biochar stability over time" and so "it is too early to rely on biochar as an effective climate mitigation tool." The 2014 Intergovernmental Panel on Climate Change report reviewing the literature on mitigation found that biochar might be able to remove substantial CO_2 from the air—if there were enough available biomass and if further research and field validation were able to verify high levels of long-term biochar stability in the soil.

The U.S. Congressional Budget Office has identified a variety of other practices that could increase the carbon stored in farmland. For instance, "as farmers rotate which crops they grow on which parts of their land from year to year, they can foster sequestration through frequent use of cover crops—particularly those, like hay, that do not require tillage and that fix carbon in the soil through their extensive root systems." Other practices that could help store more carbon in farmland include preventing erosion by "planting grasses on the edges of cropland and streams." Also, grazing management strategies, which includes grazing areas rotation and improved plant species, can help reduce carbon loss on rangeland and pasture.

Third, some crops and livestock produce lower amounts of total GHG emissions per calorie delivered than others. In particular, a December 2014 literature review by the Chatham House, the Royal Institute of International Affairs based in London, points out that "greenhouse gas emissions from the livestock sector are estimated to account for 14.5% of the global total." That means the full life-cycle GHG emissions of meat and dairy are comparable with the direct emissions from the global transport sector. Beef and dairy, which are the most emissions-intensive of livestock products, generate 65% of the total GHGs emitted by livestock. Globally, the GHG emissions from producing beef is on average more than a hundred times greater than those of soy products per unit of protein.

Although strategies to reduce those emissions can play a substantial role in reducing their emissions, shifting dietary trends could play a much bigger role. The 2014 IPCC mitigation report has a chapter on "Agriculture, Forestry and Other Land Use." It concludes that dietary changes are critical to achieving a 2°C target. The IPCC noted that studies have estimated "agricultural non-CO_2 emissions (CH_4 and N_2O) would triple by 2055 to 15.3 $GtCO_2eq/yr$ [billion tons of CO2-equivalent a year] if current dietary trends and population growth were to continue." In terms of reducing that unsustainable increase, the IPCC noted that the "decreased livestock product" scenario would achieve double the emissions reductions that "technical mitigation options on the supply side, such as improved cropland or livestock management" could achieve. The two together could bring future emissions down to 2.5 $GtCO_2eq/yr$.

The IPCC looked at studies that compared business-as-usual GHG emissions in the sector with options that included reduced meat and dairy consumption. They found that "Changed diets resulted in GHG emission savings of 34–64% compared to the 'business-as-usual' scenario." Adopting the Harvard Medical School's "healthy diet"—in which meat, fish, and egg consumption are no more than 90 grams per capita a day—"would reduce global GHG abatement costs to reach a 450 ppm CO_2eq concentration target by ~50% compared to the reference case." Of course, the authors quickly add, "Considerable cultural and social barriers against a widespread adoption of dietary changes to low-GHG food may be expected."

Dietary changes, like all behavior changes, are likely to be some of the slowest type of mitigation strategies to be adopted. Usually a major change in behavior requires a broad societal realization that the behavior is harmful to both the individual and society. In the case of GHG emissions from the food sector, it may simply be the reality of climate change that ultimately drives dietary change. If some of the business-as-usual projections discussed in this book occur, then the world will lose some one third of its most arable land to near-permanent drought and Dust-Bowlification post-2050. At the same time, acidification, along with saltwater infiltration to rich agricultural deltas from sea-level rise, will threaten more sources of food. In addition, this all happens just when we are projected to add another 3 billion people to feed. There is unlikely to be sufficient arable land and fresh water available to sustain all those people on a Western meat-intensive diet. Some combination of rising food prices and government policy and societal pressure to avoid mass starvation could well bring about dietary change.

Finally, it is precisely because the world will likely have to devote so much of its arable land and fresh water to feeding its population post-2050 that it is difficult to see how the agricultural sector will be able to devote significant resources to biofuels. Certainly, our current generation of land- and water-intensive crop-based biofuels are unlikely to be tenable. For there to be substantial bioenergy production after mid-century—other than through agriculture and food waste—would require a commercially viable next-generation biofuel to be developed that can be grown at large scale on nonagricultural land with minimal water input.

What role can energy conservation play?

Energy conservation is reducing energy consumption through behavior change. Just as dietary change on a large scale could be a very big reducer of greenhouse gas emissions, so could simply changing personal behavior, if it were done on a large scale. Energy conservation could potentially be one of the largest, if not the largest, sources of GHG reductions. However, figuring out how motivated people might be to voluntarily conserve energy in the future is exceedingly difficult, because it requires imagining how people in, say, 2030, will look at their responsibility to the future as it becomes more and more painfully clear that not changing behavior will have catastrophic impacts for billions of people, including their own children and grandchildren.

For the majority of the biggest GHG emitters in the world, especially those in the developed countries, how many of the energy-intensive activities we do every week and every year are truly vital, something we could not live without? How big a house is vital? How much driving is discretionary? How much flying? These are not questions that can easily be answered today.

The physicist Saul Griffith has calculated one of the most comprehensive personal carbon footprints ever done. He did not just examine how much energy is consumed by his commuting and his flying and the appliances in his home, he also calculated in detail the energy needed to manufacture and transport all of the stuff he uses, such as his catamaran, which he uses so much he has to replace it every couple of years. His scientific conclusion is as follows: "A quarter of the energy we use is just in our crap." How much of our crap can we live without?

As with dietary change, energy conservation is likely to be adopted relatively slowly until there is broad societal realization that carbon pollution is harmful to both the individual and society. However, there are signs that society is beginning to make that realization. For instance, in November 2014, Pope Francis wrote a letter to world leaders saying, "There are constant assaults on the natural environment, the result of unbridled consumerism, and this will have serious consequences for the world economy." In June 2015, the Pope issued his powerful 195-page Encyclical statement on climate change, which spelled out why it is the transcendent moral issue of our time. As more moral leaders speak out about the potentially catastrophic consequences of our current behaviors, more people may be willing to consider changing them.[65]

7

CLIMATE CHANGE AND YOU

This chapter will explore some of the more personal questions that climate change raises for individuals and their families.

How will climate change impact you and your family in the coming decades?

The transition to a low carbon economy is inevitable this century, and indeed it has already begun. It will have significant consequences for both you and your family, whether the transition comes fast enough to avoid dangerous warming of more than 2°C or not. As we have seen, because climate action has been so delayed for so long, humanity cannot avoid very serious climate impacts in the coming decades—impacts that will affect you and your children. Therefore, you need to understand what is coming so that you and your family will be prepared. At this point, it seems likely that climate change will transform the lives of your children more than the Internet has.

The defining story of the 21st century is a race between the impacts our cumulative carbon emissions will increasingly have on our climate system and humanity's belated but accelerating efforts to replace fossil fuels with carbon-free energy. Some of the most significant impacts of climate change are ones that we likely have not foreseen. For instance, a couple of decades ago, few people imagined that the most consequential near-term impacts of climate change on large parts of both the United States and Canada would be the warming-driven population explosion of a tiny pest, the tree-destroying bark beetle. Only through a comprehensive and ongoing under-standing of climate impacts and the clean energy transition will you be able to determine how climate change might affect your family. This chapter will explore (1) a few of the bigger and more obvious consequences that could affect a large frac-tion of the people reading this book as well as (2) some choices you may face in the coming years.

How might climate change affect the future price of coastal property?

Because of climate change, coastal property values in the United States and around the developed world are all but certain to crash. The latest science suggests this could well happen sooner rather than later. Such an event would have a profound impact on the local, national, and global economy. In the United States alone, at least $1.4 trillion in property lies within 660 feet of the U.S. coast, as a detailed analysis by Reuters found. Worse, "Incomplete data for some areas means the actual total is probably *much* higher." Globally, coastal property is worth many times that.[66]

In 2014 and 2015, a wealth of observations and analyses revealed that large parts of the great ice sheets of Antarctica and Greenland are unstable and headed to irreversible col-lapse, and some parts may have already passed the point of no return (see Chapter Three). Another 2015 study found that global sea-level rise since 1990 has been speeding up even faster than we knew. Other studies have found that the U.S. East Coast in particular is experiencing faster sea-level rise than the rest of the world, and this is a trend that could well continue for the rest of the century.

The recent findings have led many experts to revise their sea-level rise prediction upward and conclude that we are headed toward what used to be the high-end of projected global sea-level rise this century, 4–6 feet or more. The conse-quences of such sea-level rise to low-lying developing countries will be catastrophic, as tens of millions will be forced to abandon their homes and move inland. It will also be catastrophic to the developed countries. A 2013 National Oceanic and Atmospheric Administration study found that, under the kind of rapid sea-level rise scientists now worry about, the New Jersey shore from Atlantic City south would see Sandy-level storm surges almost every year by mid-century.

And this is not even the current plausible worst-case scenario any more. In July 2015, a number of top scientists, led by James Hansen, one of America's most renowned and prescient climatologists, warned that failure to curb carbon pollution rapidly could even lead to 10 feet of sea level rise by 2100.

Informed people such as John Van Leer, an oceanographer at the University of Miami, worry that one day, they will no longer be able to insure or sell their houses. "If buyers can't insure it, they can't get a mortgage on it. And if they can't get a mortgage, you can only sell to cash buyers," Van Leer says. That is particularly true in a place such as Miami where, as discussed, conventional efforts to deal with sea-level rise, such as sea walls and barriers, will not work because South Florida sits above a vast and porous limestone plateau.

Despite the reality of accelerating sea level rise, coastal properties are soaring in many places. One of the most vulnerable spots in the developed world to both warming-driven sea-level rise and storm surge is Miami. However, between the first quarter of 2013 and the same period in 2014, Miami property values were up 19%. The sales price for luxury homes—the top 10% of the market—jumped even more, 34%. By itself, the United States appears to be in a coastal real estate bubble of over 1 trillion dollars. Florida is ground zero because it leads the country with $484 billion in "property covered by the National Flood Insurance Program, often at below market." The Miami area is so flat that even with a mere 3 feet of sea-level rise, "more than a third of southern Florida will vanish; at six feet, more than half will be gone." For all these reasons, Harold Wanless, chair of University of Miami's geological sciences department, said in 2013, "I cannot envision southeastern Florida having many people at the end of this century." In 2014, he said, "Miami, as we know it today, is doomed. It's not a question of if. It's a question of when." Yet Chuck Watson, a disaster-impact analyst with a great deal of Florida experience, has pointed out: "There is no serious thinking, no serious planning, about any of this going on at the state level. The view is, 'Well, if it gets real bad, the federal government will bail us out.' It is beyond denial; it is flat-out delusional."

The federal government and American taxpayers may well be willing to bail out Florida if a major hurricane hits Miami or Tampa in the next decade or two. However, it seems unlikely that they will continue to do so repeatedly as it becomes increasingly obvious that seas are going to keep rising and devastating storm surges are going to become commonplace. At some point, then, it will be all but impossible to get flood insurance, and so the only people who could afford to buy coastal property will be cash-buyers wealthy enough to lose their entire investment in the next storm.

Coastal property values will almost certainly collapse once homes become uninsurable or when governments decide that spending money to constantly rebuild or "nourish" coastlines makes no sense. For instance, currently many communities on the East Coast dealing with sea-level rise and storm surges nourish their beaches and slow down erosion by strengthening them with large amounts of sand. The Federal government currently covers on average about two thirds of the cost. A March 2015 journal article concludes "a sudden removal of federal nourishment subsidies, as has been proposed, could trigger a dramatic downward adjustment in coastal real estate, analogous to the bursting of a bubble." Their model suggests that sudden removal would cause property values to decline 17% to 34% if it happened today, and a much more dramatic decline will occur if that seemingly inevitable policy change were to happen in a decade or two.

So, when will coastal property values crash? The exact time is unknowable since it might be triggered by New York and New Jersey, or South Florida or New Orleans, being hit by a Sandy-like storm surge in the next decade or so. However, the crash does not have to wait for seas to engulf an area. It could occur when the smart money figures out we have dawdled too long to save the coast from rapid sea-level rise and constant devastating storm surges. We have not hit that critical mass of knowledge yet. If we had, construction and property values certainly would not be soaring in so many coastal areas. However, the attention given recent observations of Antarctic and Greenland instability—with the *New York Times* pointing out in 2014 that this could lead to "enough sea-level rise that many of the world's coastal cities would eventually have to be abandoned"—suggests we are closer to the tipping point than people realize.

What does all this mean for you? A peak in coastal property in the next decade or two seems plausible. Certainly buying coastal property today as a long-term investment seems to fly in the face of science. Whether it makes sense to keep coastal property you already own is a decision that can be made only at a personal level. However, if your family is planning on selling the property someday, you will want to beat the bursting of the bubble, whenever you imagine that is going to occur.

How might climate change affect decisions about where to live and retire in the coming decades?

A great many people who retire, especially in the developed countries, do so to places that are warm or near the coast or both. It already seems clear that coastal property is probably not a wise investment for someone planning to retire in the next decade or two, unless you are so wealthy can afford to lose your entire investment. Many retirees choose a relatively warm and dry climate, such as the Mediterranean or U.S. southwest. Coastal Mediterranean will eventually not be an option thanks to sea-level rise. However, as we have seen, virtually all of the warm semiarid climates in the world—what are typically labeled the subtropics—are going to become hot and arid. They are going to Dust-Bowlify, and water will become much scarcer and more expensive, perhaps leading to rationing. That is particularly true if we do not get on the 2°C path quickly.

Ultimately, the best places to live are ones that are neither coastal nor semiarid nor already hot today. Because so many places in the world will eventually have problems providing enough food or water to their inhabitants, places where there is relatively abundant water and arable land would seem to be the best choice for where to retire. Those places will also be the among the most desirous places to live by mid-century. This includes the northern mid-west United States and northern Europe and the like. Presumably, smart investors will figure this out in the next decade or two. That said, there are no locations that qualify as "winners" in a changing climate. Many may have thought Russia would see nothing but benefit from global warming. The devastating heat wave that hit in the summer of 2010 made clear that was not true, as tens of thousands of Russians died and the country was forced to stop all grain exports for a year, causing food prices to skyrocket.

I do not think anyone can say exactly when the population movement in, say, the United States reverses from its current north-to-south direction, or when soaring food prices become commonplace. However, these changes are inevitable, and the people who plan ahead for that outcome will come out ahead.

What should students study today if they want to prepare themselves for working in a globally warmed world?

In the coming decades, more and more money and resources and people will be devoted toward (1) adapting to whatever climate change we fail to stop and (2) stopping climate change from getting even worse. Climate change and our response to it will create trillion-dollar industries in low-carbon energy, energy efficiency, sustainable agriculture, and every type of adaptation imaginable. Students who want to be employable in a carbon-constrained world while contributing to the solution will have a great many choices and options available to them, if they start studying and planning now.

The transition to a more efficient, low-carbon economy is inevitable because of the reality of climate change. The question has only been "How fast?" Because of the significant (and growing) commitments to clean energy by China, the European Union, most U.S. states, and many other countries, the transition has been jumpstarted. Investment in clean energy is already a few hundred billion dollars a year, and over the next decade or so it should hit $1 trillion or year, and then double again in the next decade or two after that. Therefore, there will be a great need for engineers and researchers and entrepreneurs of all type. There will also be a great need for people with specific expertise in solar power and wind power, energy storage, electric cars, and energy efficiency in every sector, from buildings, to industry, to transportation. These projects will need financing and legal contracts and the like. Many will need architects and urban planners. Thus, simply being well schooled and experienced in, say, sustainable architecture or green financing will be valuable.

Although the low-carbon transition is now irreversible and accelerating, it is still is a long way from stopping catastrophic warming, let alone stopping dangerous warming. Thus, there will be similarly large investments made in critical aspects of adaptation. Providing food (and water) for 9 billion or more people by mid-century in a world of rapid climate change is going to be the greatest challenge humanity has ever faced. Therefore, we will need experts in sustainable agriculture, marine biology, agronomy, hydrology, and on and on. The combination of ever-rising seas and storm surges means that all of our major coastal areas that are not abandoned will need to be protected. Thus, the sea wall and levee business will boom. Because we are looking at a rate of sea-level rise that could approach one foot a decade by century's end, ports and other key coastal infrastructure will have to be completely reimagined. It is going to get hot, so the air conditioning business will boom.

I advise students to learn as much as they can about climate science and climate solutions, while they are figuring out what type of career area they are interested in and have talent for (i.e., science, engineering, law, design, medicine and health care, media, and so on). Then, I encourage them to find the intersection of those two areas. For example, a doctor could become an expert in tropical diseases, since many are not going to be purely tropical anymore, or she could become an expert in heat-related illnesses. Moreover, a doctor could go into urology, since a hotter world means more dehydration and more kidney stones. The point is, the entire world is going to be transformed in ways both easy and hard to imagine. The more you know, and the more you apply your imagination to what you know, the more employable and adaptable you will be.

Should climate change affect how you invest for the future?

It already seems clear that certain investments are riskier than others and getting more so every passing year. A clear example is coastal property. Also, in Chapter Six, I discussed a study that found "globally, a third of oil reserves, half of gas reserves and over 80% of current coal reserves should remain unused from 2010 to 2050 in order to meet the target of 2°C." It is entirely possible that we will use more of our fossil fuel resources than that and exceed 2°C. However, it seems very unlikely that we would burn all of our known fossil fuel reserves, since that would mean warming beyond 6°C (11°F) and an irreversibly ruined climate that would have trouble sustaining 1 billion people. It would also mean that (1) the Earth ends up with very large areas that are uninhabitable or unfarmable and (2) the ocean would suffer a mass extinction, with large and growing dead zones.

If you think humanity is going to be smart enough to leave a large fraction of our fossil fuel resources in the ground, then that would mean trillions of dollars worth of reserves will become essentially valueless, probably sometime in the next couple of decades. In that case, it seems inevitable that fossil fuel companies and companies that service the industry are overvalued. They would be in a bubble of their own that must burst. Should that affect your decision on how to invest? More and more major institutions in the United States and around the world are divesting from fossil fuel-related investments, either because they think those investments are "wrong" or because they think their bubble must burst or both. In addition, more and more financial institutions and advisors are offering portfolios for individuals who want to divest of this risk.

In terms of investing in the possible "winners," that has its own risks. A company could be in exactly the right business area (i.e., solar power, sea walls, or drought-resistant crops) but still be poorly managed and go bankrupt—or a competitor could have a better product. Thus, there are no recommendations here for specific companies. However, if making wise (and sensibly diversified) investments is important to you, then being highly informed on climate science and solutions will certainly give you an edge.

How can you reduce your carbon footprint?

You may decide that you would like to reduce your impact on the climate—either because you think it is the right thing to do or because you want some experience in an inevitable transition that ultimately the vast majority of people will be making. Here is a brief discussion of the most important things that you can do now and in the near future to reduce your family's carbon footprint, the total greenhouse gas (GHG) emissions released as a result of your purchases and choices.

The biggest contributors to your carbon footprint are your home, your transportation, your stuff, and your diet (which is discussed separately below). As of today, perhaps the single biggest thing you can do to reduce your home's carbon footprint is get a solar panel installed on the roof. In a growing number of locations, companies will do that without your having to put any money up front. In that scenario, you will lease the solar panel—and increasingly the deal is that your monthly lease is guaranteed to be lower than the cost of the electricity you had been purchasing. Therefore, it is a good deal if this option is available in your area, although it may be that other financing options make more sense depending on where you live and what incentives are available to you. If you cannot install your own solar system, then you should find out whether your local utility or other service provider will sell you zero-carbon electricity, such as from new wind turbines. You should also make your home more energy efficient. Most utilities offer energy audits that identify your biggest opportunities for savings, and many utilities offer rebates and other incentives for more efficient lighting and other appliances.

In terms of your transportation emissions, the most basic thing that you can do is not travel by air as much. If you or your family are traveling by air more than once a year, then that may well be the biggest single contributor to your carbon footprint. Vacation spots you drive to are almost certainly much less carbon intensive than those you fly to. Trains are best of all. If you do fly a lot, you probably know that you can ostensibly "offset" those emissions, since the option is offered by many travel websites and airlines. However, many of these so-called offsets simply represent payments to existing clean energy facilities, rather than money going to fund new low-carbon infrastructure. Thus, if the offsets are cheap, they probably do not actually offset a lot of new emissions.

The other key piece of reducing your transportation footprint is your personal vehicle. You will certainly save a lot of carbon if you can telecommute part-time, use mass transit, or ride a bike. For the driving you cannot avoid, the primary thing you can do today is purchase the most fuel-efficient or high-mileage car that will suit your needs. The Toyota Prius, which is the car I have driven for 10 years, remains one of the most efficient vehicles ever designed in its class. Other hybrids and even modern diesel cars are also worth considering.

Ultimately, if you want to reduce your vehicular emissions, then you will want to look into the growing array of plug-in hybrid electric vehicles and pure electric vehicles (see Chapter Six). That is particularly true if you live in a state like California or a country like Denmark, where the electricity supply generates far less carbon emissions than other states and countries. Over the next few years, as the price of batteries continues to drop and more and more major car companies offer electric vehicles with a 200-mile range and relatively fast charging, this option will become more and more attractive. By 2020, the combination of an electric car with home solar panels (and possibly home storage batteries) is likely to be a cost-effective zero-emissions solution.

Hydrogen fuel cell cars are unlikely to be a cost-effective and practical way to reduce your vehicular GHG emissions for a long time, if ever. That would require a number of technology breakthroughs, as Ford Motor Company has said. It will also require there to be hundreds (if not thousands) of fueling stations that provide affordable carbon-free hydrogen—and not hydrogen from natural gas. Finally, it would require that there is no electric vehicle that will meet your needs, because electric cars travel approximately three times as far on a given amount of carbon-free electricity as hydrogen fuel cell vehicles.

Finally, there is the carbon embedded in all of the stuff that you buy, the carbon dioxide released in producing the materials stuff is made out of, in manufacturing your stuff, and then in transporting it. As noted in the last chapter, physicist Saul Griffith calculated "A quarter of the energy we use is just in our crap." Ultimately, everything you purchase adds to your current footprint. In general, the more material it is made out of and the more expensive it is, the more GHGs were released in making it and getting it to you. Therefore, if you want to trim your family's carbon footprint, remember the motto, "small is beautiful."

What role can dietary changes play in reducing your carbon footprint?

If you have a diet rich in animal protein, then it is likely you can significantly reduce your greenhouse gas emissions by replacing some or all of that with plant-based food. That is particularly true if your diet is heavy in the most carbon intensive of the animal proteins, which includes lamb and beef but also dairy. Globally, the GHG emissions from producing beef is on average more than a hundred times greater than those of soy products per unit of protein.[67]

According to the world's top scientists in their 2014 survey of the scientific literature, various world diets that reduced meat and dairy consumption, "resulted in GHG emission savings of 34–64% compared to the 'business-as-usual' scenario." Also, if the world adopted the Harvard Medical School's "healthy diet," in which meat, fish, and egg consumption are no more than 90 grams per capita a day, the cost of avoiding catastrophic warming would be cut by 50%.

As with the other ways to reduce your carbon footprint, this is of course a matter of choice for you and your family. Also, many studies point to diets lower in meat as healthier. As discussed earlier, if some of the business-as-usual projections discussed in this book occur, then the world will lose one third of its most arable land to near-permanent drought and Dust-Bowlification post-2050. At the same time, acidification and rising saltwater infiltration of rich agricultural deltas will threaten more sources of food. In such a ruined climate, we still have to figure out how to feed another 3 billion people. There is unlikely to be sufficient arable land and fresh water available to sustain all those people on a Western meat-intensive diet. Some combination of rising food prices and government policy and societal pressure to avoid mass starvation could well bring about dietary change.

What is the best way to talk to someone who does not accept the growing body of evidence on climate science?

Having read this book, you now know more about climate change than most people you will meet, unless you work at a place like NASA or the International Energy Agency. Because there is a growing national and global conversation on climate change, with major world figures like the Pope joining in, you are likely to encounter people who do not know basic climate science or actually "know" things that are not true. In particular, certain flawed arguments against the science of human-caused climate change have become very commonplace. These myths have become popular for two key reasons. First, most of them are repeated again and again by the disinformation campaign discussed in Chapter Five. Second, they sound plausible on the surface.

Anyone who plans to talk about climate change with their friends or family or colleagues should spend some time at the website SkepticalScience.com. Skeptical Science tracks and debunks the most popular climate science myths. It provides both simple and more detailed responses to all of the myths, complete with detailed citations of and links to the recent scientific literature. It even has an app for that purpose. Furthermore, it also includes the best strategies for effective communications based on the social science literature. By permission, I will make use of their material below—with tweaks and additions—to provide short answers to the myths and questions you are most likely to hear (which are in quotation marks).

1. "The climate has changed before" or "The climate is always changing." This assertion is actually true, but it is meant to imply that because the climate changed before humans were around, humans cannot cause climate change. That is a logical fallacy, like saying smoking cannot cause lung cancer because people who do not smoke also get lung cancer. In fact, climate scientists now have the same degree of certainty that human-caused emissions are changing the climate as they do that cigarette smoking is harmful. The key point is that the climate changes when it is forced to change. Scientific analysis of past climates shows that greenhouse gases, principally CO_2, have controlled most ancient climate changes. The evidence for that is spread throughout the geological record. Now humans are forcing the climate to change far more rapidly than it did in the past mainly by our CO_2 emissions—50 times faster than it changed during the relatively stable climate of the past several thousand years that made modern civilization (and particularly modern agriculture) possible.

2. "Warming has stopped, paused, or slowed down." In fact, 2014 was the hottest year on record (and 2015 is on track to be even hotter). The warming trend in the past two decades is nearly identical to the warming trend in the two decades before that. Also, empirical measurements of the Earth's heat content show the planet is still accumulating heat. Global warming is still happening everywhere we look, especially the oceans, where more than 90% of the extra heat trapped by human carbon pollution goes.

3. "There is no scientific consensus on human-caused warming": In fact, our understanding that humans are causing global warming is the position of the Academies of Science from 80 countries plus many scientific organizations that study climate science. More specifically, surveys of the peer-reviewed scientific literature and the opinions of experts consistently show a 97%–98% consensus that humans are causing global warming.

4. "Recent warming is due to the sun." In fact, in the last 35 years of global warming, the sun and the climate have been going in opposite directions—with the sun actually showing a slight cooling trend. The Sun can explain some of the increase in global temperatures in the past century, but a relatively small amount. The best estimate from the world's top scientists is that humans are responsible for *all* of the warming we have experienced since 1950.

5. "Climate change won't be bad." As the scientific literature detailed in this book makes clear, the negative impacts of global warming on agriculture, the environment, and public health far outweigh any positives. The consequences of climate change become increasingly bad after each additional degree of warming, with the consequences of 2°C being quite damaging and the consequences of 4°C being catastrophic. The consequences of 6°C would be almost unimaginable.

6. "Can climate models be trusted?" A related question is, "Since we can't predict the weather a few weeks from now, how can we predict the climate a few decades from now?" Although there are uncertainties with climate models, they successfully reproduce the past and have made predictions that have been subsequently confirmed by observations. Long-term weather prediction is hard because on any given day a few months from now or a few years from now, the temperature could vary by tens of degrees Fahrenheit or even Celsius. Similarly, there could be a deluge or no rain at all on any given day. The weather is the atmospheric conditions you experience at a specific time and place. Is it hot or cold? Is it raining or dry? Is it sunny or cloudy? The climate is the statistical average of these weather conditions over a long period of time, typically decades. Is it a tropic climate or a polar climate? Is it a rainforest or a desert? The climate is considerably easier to predict precisely because it is a long-term average. Greenland is going to

be much colder than Kenya during the course of a year and during almost every individual month. The Amazon is going to be much wetter than the Sahara desert virtually year-round.

7. "Are surface temperature records reliable?" Independent studies using different software, different methods, and different data sets yield similar results. The increase in temperatures since 1975 is a consistent feature of all reconstructions. This increase cannot be explained as an artifact of the adjustment process, the use of fewer temperature stations, or other nonclimatological factors. Natural temperature measurements also confirm the general accuracy of the instrumental temperature record.

8. "Isn't Antarctica gaining ice?" Satellites measurements reveal Antarctica is *losing* land ice at an accelerating rate, leading many scientists to increase their projections of sea-level rise this century. Why, then, is Antarctic *sea ice* growing despite a strongly warming Southern Ocean? The U.S. National Snow and Ice Data Center explained in 2014 that the best explanation from their scientists is that it "might be caused by changing wind patterns or recent ice sheet melt from warmer, deep ocean water reaching the coastline. ... The melt water freshens and cools the deep ocean layer, and it contributes to a cold surface layer surrounding Antarctica, creating conditions that favor ice growth."[68]

9. "Didn't scientists predict an ice age in the 70s?" The 1970s ice age predictions you hear about today were predominantly from a very small number of articles in the popular media. The majority of peer-reviewed research at the time predicted warming due to increasing CO_2.

Do we still have time to preserve a livable climate?

There is definitely time to avert the worst impacts of climate change, and I personally have become more optimistic of humanity's chances in the last year. Yes, if you focus only on the latest climate science and the inadequacy of our political leadership, then you can to fall into despair and pessimism—and perhaps even denial. Even for pessimists, though, a key point of this book is that trying to live your life without thinking about climate change is a losing strategy for you and your family. The more you know, the better you will be able to plan for the future.

But recently there have been many hopeful signs. For instance, in the year leading up to the December 2015 Paris climate talks, the leading nations of the world have made public commitments to reverse their unsustainable greenhouse gas emissions trends. My June 2015 trip to China to meet with top governmental and non-governmental experts on clean energy and climate made clear to me the country's leaders are serious about cleaning up their polluted air, beating their climate targets, and deploying carbon-free energy even more rapidly. It will still take considerably more effort to keep total warming below the 2°C defense line that top scientists increasingly tell us we must not cross. But we have collectively started to take actions needed to keep that possibility alive, albeit barely.

Another hopeful sign is that the key technologies needed to avert catastrophic warming—solar power, wind, energy-efficient lighting, advanced batteries—have seen a steady and in some cases remarkable drop in prices. This price drop has been matched by a steady improvement in performance. Maybe at some point in the past you could believe that climate action was too expensive, but not any more. The nation's top scientists, energy experts, and governments have all spelled out in great detail that even the strongest climate action is super cheap.

Finally, we have seen more and more opinion makers speak out on climate change. Perhaps the most significant among them is Pope Francis, whose 195-page encyclical in June 2015 has spurred a global debate about the moral urgency for climate action. I would urge anyone needing motivation to accept and tackle the challenges we face in the years ahead to read it. The Pope's message is at its core a simple one: "We must regain the conviction that we need one another, that we have a shared responsibility for others and the world, and that being good and decent are worth it."

PRIMARY SOURCES

American Association for the Advancement of Science Climate Science
Panel. (2014). What we know. Retrieved from whatweknow.aaas.
org.

International Energy Agency. (2014, September). Capturing the multiple benefits of energy efficiency. Retrieved from iea.org/bookshop.

IEA *Technology Roadmap* series, *Geothermal Heat and Power* (2011),
Biofuels for Transport (2011), *Bioenergy for Heat and Power* (2012),
Hydropower (2012), *Wind Energy* (2013), *Solar Photovoltaic Energy*
(2014), *Solar Thermal Electricity* (2014), *Nuclear Energy* (2015), with
Nuclear Energy Agency. Retrieved from iea.org/publications.

New, M., et al. editors. (2011, January). Theme issue, four degrees and
beyond: the potential for a global temperature increase of four
degrees and its implications. *Philosophical Transactions of the Royal
Society A*. Retrieved from rsta.royalsocietypublishing.org/content/369/1934/6.full.

Potsdam Institute for Climate Impact Research and Climate Analytics.
(2012, November). Turn down the heat: why a 4°C warmer world
must be avoided. Report for the World Bank. Retrieved from www.
worldbank.org/en/topic.

UK Met Office Hadley Centre. (2014). Climate risk: an update on the
science. Retrieved from www.metoffice.gov.uk/media/pdf/o/p/
Climate_risk_an_update_on_the_science.pdf.

U.S. National Academy of Sciences and British Royal Society. (2014).
Climate change: evidence & causes. Retrieved from royalsociety.
org/policy/projects/climate-evidence-causes.

U.N. Intergovernmental Panel On Climate Change Fifth Assessment
Report, Working Group I. *The Physical Science Basis*, 2013, Working
Group II, *Impacts, Adaptation and Vulnerability*, 2014, Working Group
III, *Mitigation of Climate Change*, 2014, and *Synthesis Report*, 2014. All
references to these reports are to their Summary for Policymakers
unless otherwise stated in the text. Retrieved from www.ipcc.ch.

U.S. Global Change Research Program. (2014). Climate Change Impacts
in the United States: The Third National Climate Assessment.
Retrieved from nca2014.globalchange.gov.

Vatican (2015). Encyclical Letter *Laudato si'* of the Holy Father Francis
on care for our common home. Retrieved from w2.vatican.va/
content/vatican/en.html.

NOTE: References that are identified in the text (by, for example, the
title of an article and the journal it appears in) will generally not be
repeated in the footnotes.

NOTES

1. Thompson, L. (2010, Fall). Climate change: the evidence and our options. *Behavior Analyst*. Retrieved from http://researchnews.osu.edu/archive/TBA--LTonly.pdf.
2. Jevrejeva, S., et al. (2014). Upper limit for sea level projections by 2100. *Environmental Research Letters*, Vol. 9; Shepherd, A., et al. (2013). A reconciled estimate of ice-sheet mass balance. *Science*; NASA *Science News*. (2012, July 24). Satellites see unprecedented greenland ice sheet surface melt; Hickey, H., & Ferreira, B. (2014, February). Greenland's fastest glacier sets new speed record. *UWToday*; Alfred Wegener Institute news release. (2014, August). Record decline of ice sheets; McMillan, M., et al. (2014, June). Increased ice losses from Antarctica detected by CryoSat-2. *Geophysical Research Letters*; NASA news release. (2014, May 12). NASA-UCI study indicates loss of West Antarctic glaciers appears unstoppable; American Geophysical Union news release. (2014, December). West Antarctic melt rate has tripled.
3. Balmaseda, M., et al. (2013, May). Distinctive climate signals in reanalysis of global ocean heat content. *Geophysical Research Letters*; Durack, P., et al. (2014). Quantifying underestimates of long-term upper-ocean warming. *Nature Climate Change*.

4. Cook, J. 10 Indicators of a human fingerprint on climate change. www.skepticalscience.com/10-Indicators-of-a-Human-Fingerprint-on-Climate-Change.html. (This link contains hyperlinks to references for all of the indicators.)

5. Broecker, W., (1995, July). Cooling the Tropics. *Nature*.

6. Dessler, A., et al. (2008, October). Water-vapor climate feedback inferred from climate fluctuations, 2003–2008. Geophysical Research Letters; Tripati, A. (2009). Coupling of CO2 and ice sheet stability over major climate transitions of the last 20 million years. *Science*; Zeebe, R., & Caldeira, K. (2008). Close mass balance of long-term carbon fluxes from ice-core CO2 and ocean chemistry records. *Nature Geoscience*.

7. National Science Foundation Press Release. (2013, March). Earth is warmer today than during 70 to 80 percent of the past 11,300 years.

8. Conway, E. (2008, December). What's in a name? Global warming vs. climate change. NASA.

9. U.S. EPA. (2014). Greenhouse gas emissions. www.epa.gov/climatechange/ghgemissions.

10. Rahmstorf, S. (2013, November). Global warming since 1997 underestimated by half. www.realclimate.org.

11. Foster, G., & Rahmstorf, S. (2011, December). Global temperature evolution 1979–2010. *Environmental Research Letters*; Tollefson, J. (2013, August). Tropical ocean key to global warming 'hiatus'. *Nature News*.

12. Solomon, S. et al. (2009). Irreversible climate change due to carbon dioxide emissions. *Proceedings of the National Academy of Sciences*; Hickey, H. (2014, May). West Antarctic ice sheet collapse is under way. *UWToday*; Rahmstorf, S. (2013). Sea-level rise: where we stand at the start of 2013. www.realclimate.org.

13. Rupp, D., et al. (2012). Did human influence on climate make the 2011 Texas drought more probable? *Bulletin of the American*

Meteorological Society; National Center for Atmospheric Research. Record high temperatures far outpace record lows across U.S. *AtmosNews*, November 2009; UK Met Office Hadley Centre. Climate risk: an update on the science, 2014.

14. Masters, J. (2011, June). 2010–2011: Earth's most extreme weather since 1816? *Weather Underground Blog*. Retrieved from www. wunderground.com/blog/JeffMasters; National Weather Service news release. (2010, May 18) May 1 & 2 2010 Epic Flood Event for Western and Middle Tennessee. Retrieved from www.srh.noaa. gov/ohx/?n=may2010epicfloodevent.

15. Sallenger, A. Jr., et al. (2012). Hotspot of accelerated sea-level rise on the Atlantic coast of North America. *Nature Climate Change*.

16. Rahmstorf, S., & Coumou, D. Extremely hot, March 26, 2012. Retrieved from RealClimate.org; Hansen, J., et al. (2012, August). The new climate dice: public perception of climate change. NASA News Brief. Retrieved from www.giss.nasa.gov/research/briefs/ hansen_17.

17. Griffin, D., & Anchukaitis, K. (2014, December). How unusual is the 2012-2014 California drought? *Geophysical Research Letters*; U.S. Geological Survey news release. (2011, June 9). USGS study finds recent snowpack declines in the Rocky Mountains unusual compared to past few centuries; U.S. Geological Survey news release. (2013, May 13). Warmer springs causing loss of snow cover throughout the Rocky Mountains.

18. Vose, J., et al. (Eds.) (2012). Effects of climatic variability and change on forest ecosystems: a comprehensive science synthesis for the U.S. forest sector. U.S. Forest Service.

19. National Climate Assessment charts. Retrieved from nca2014. globalchange.gov/downloads.

20. Changnon, S., et al. (2006, August). Temporal and spatial characteristics of snowstorms in the contiguous United States. *Journal of Applied Meteorology and Climatology*; Madsen, T., & Willcox, N.

(2012, Summer). When it rains, it pours: global warming and the increase in extreme precipitation from 1948 to 2011. Environment America Research & Policy Center; National Oceanic and Atmospheric Administration chart. Retrieved from www.ncdc.noaa. gov/extremes/cei/graph/ne/4/10-03; Alfred Wegener Institute news release. (2012, January). New study shows correlation between summer Arctic sea ice cover and winter weather in Central Europe; Radford, T. (2014, September) Less snow under global warming may not halt blizzard hazard. *Scientific American*.

21. Emanuel, K. (2015, March 18). Severe tropical cyclone Pam and climate change. Retrieved from www.realclimate.org; Grinsted, A., et al. (2012). Homogeneous record of Atlantic hurricane surge threat since 1923. *Proceedings of the National Academy of Sciences*; Graumann, A., et al. (2005, October). Hurricane Katrina, a climatological perspective. National Climatic Data Center. (Updated August 2006.); Grinsted, A. et al. (2013, April). Projected Atlantic hurricane surge threat from rising temperatures. *Proceedings of the National Academy of Sciences*.

22. Satellite data reveal the rapid darkening of the Arctic. (2014, February 17). Scripps News.

23. Francis, J., & Vavrus, S. (2015, January). Evidence for a wavier jet stream in response to rapid Arctic warming. *Environmental Research Letters*; Duke Environment News. (2010, October 27). Increasingly variable summer rainfall in southeast linked to climate change; Morello, L. (2011, December 8). NOAA chief calls storm-ridden 2011 'a harbinger of things to come'. *ClimateWire*; Munich Re press release. (2012, October 17). North America most affected by increase in weather-related natural catastrophes; National Oceanic and Atmospheric Administration news release. (2012, October 10). Arctic summer wind shift could affect sea ice loss and U.S./European weather, says NOAA-led study; Jet Stream. Wikipedia entry; Potsdam Institute for Climate Impact Research

news release. (2014, August 12). Trapped atmospheric waves triggered more weather extremes.

24. National Oceanic and Atmospheric Administration news. (Updated 2012, March 20) 2011 tornado information; Markowski, P., & Brooks, H. (2013, December 5). Letter to the editor. Global warming and tornado intensity. *New York Times*; Elish, J. Florida State news release. (2013, September 5). Researchers develop model to correct tornado records; Oskin, B. (2013, December 11). Stronger tornadoes may be menacing US. Retrieved from LiveScience.com.

25. National Science Foundation News. (2013, May 9). Climate record from bottom of Russian lake shows Arctic was warmer millions of years ago; UCLA Newsroom. (2009, October 8). Last time carbon dioxide levels were this high: 15 million years ago, scientists report; Sluijs, A., et al. (2006, June). Subtropical Arctic Ocean temperatures during the Palaeocene/Eocene thermal maximum. *Nature*.

26. Pearce, F. (2005, August 11). Climate warning as Siberia melts. *New Scientist*; National Center for Atmospheric Research news release. (2005, December 19). Most of Arctic's near-surface permafrost may thaw by 2100; Schaefer, K., et al. (2011, April). Amount and timing of permafrost carbon release in response to climate warming. *Tellus B*; National Snow and Ice Data Center newsroom. (2011, February 16). Thawing permafrost will accelerate global warming in decades to come, says new study.

27. Kelly, R. (2013, August). Recent burning of boreal forests exceeds fire regime limits of the past 10,000 years. *Proceedings of the National Academy of Sciences*; University of Guelph news release. (2015, January 6). Peat fires—a legacy of carbon up in smoke: study; Page, S. (2002, November). The amount of carbon released from peat and forest fires in Indonesia during 1997. *Nature*; Tan, K. (2014, November, 19). Burning an ecological treasure to extinction.

Jakarta Post; NASA news release. (2012, August 28). Record temperatures and wildfires in Eastern Russia; University of Guelph news release. (2011, November 1). Drying intensifying wildfires, carbon release ninefold, Study Finds.

28. World Meteorological Organization. (2014, September). Record greenhouse gas levels impact atmosphere and oceans; Booth, B., et al. (2012, April). High sensitivity of future global warming to land carbon cycle processes. *Environmental Research Letters*; Gruber, N. (2011, April) Warming up, turning sour, losing breath: ocean biogeochemistry under global change. *Royal Society Philosophical Transactions A.*

29. Jevrejeva, S., et al. (2014, October). Upper limit for sea level projections by 2100. *Environmental Research Letters*; Bamber, J. L., & Aspinall, W. P. (2013, January). An expert judgement assessment of future sea level rise from the ice sheets. *Nature Climate Change*; Hickey, H. (2014, May). West Antarctic ice sheet collapse is under way. *UWToday*; University Of California, Irvine news release. (2014, May 18). Greenland will be far greater contributor to sea rise than expected; Helm, V., et al. (2014, August). Elevation and elevation change of Greenland and Antarctica derived from CryoSat-2. *The Cryosphere*; Geggel, L. (2015, March 18). Hidden channels beneath East Antarctica could cause massive melt. Retrieved from LiveScience.com; Rahmstorf, S. (2015, March 23). What's going on in the North Atlantic? Retrieved from RealClimate.org; Harvard University news release. (2015, January 14). Correcting estimates of sea level rise; Neumann, B., et al. (2015, March 2015). Future coastal population growth and exposure to sea-level rise and coastal flooding—a global assessment. *PLoS One*; Folger, T. Rising seas. (2013, September). *National Geographic*; Goodell, J. (2013, June 20). Goodbye, Miami. *Rolling Stone*; World Bank News. (2015, February 17). Salinity intrusion in a changing climate scenario will hit

coastal Bangladesh hard; Friedman, T. (2013, July 6). Can Egypt pull together? *New York Times*.

30. Henson, B. (2012, August 6). Dry and dryer. *AtmosNews*; NASA press release. (2015, February 12) NASA study finds carbon emissions could dramatically increase risk of U.S. megadroughts; Kahn, B. (2015, February 12). Southwest, Central Plains face 'unprecedented' drought. Retrieved from ClimateCentral.org.

31. University of College London News. (2009, May 14). Climate change: the biggest global-health threat of the 21st century; UK Met Office. (2008, December 5). Climate scientists' warning on air quality; National Center for Atmospheric Research. (2014, May 5). Climate change threatens to worsen U.S. ozone pollution. *AtmosNews*.

32. Hsiang, S. (2011, August 6, 2011). Temperature and worker output. Retrieved from www.fight-entropy.com; Dunne, J., et al. (2013, February). Reductions in labour capacity from heat stress under climate warming. *Nature Climate Change*; Gelman, A. (2012, September 17). 2% per degree Celsius … the magic number for how worker productivity responds to warm/hot temperatures. Retrieved from AndrewGelman.com; Hesterman, D. (2011, June 6). Stanford climate scientists forecast permanently hotter summers beginning in 20 years. *Stanford News*; Hsiang, S. (2012, August 21). Two percent per degree Celsius. Retrieved from www. fight-entropy.com; Rosenthal, E. (2012, August 18). The cost of cool. *New York Times*.

33. Fisk, W., et al. (2013, March). Is CO_2 an indoor pollutant? Higher levels of CO_2 may diminish decision making performance. Lawrence Berkeley National Laboratory; Satish, U., et al. (2012, December). Is CO_2 an indoor pollutant? Direct effects of low-to-moderate CO_2 concentrations on human decision-making performance. *Environmental Health Perspectives*; Kajtár, L., & Herczeg, L. (2012, April–June). Influence of carbon-dioxide

concentration on human well-being and intensity of mental work. *IDOJARAS Quarterly Journal of the Hungarian Meteorological Service*; Maddalena, R., et al. (2014, September). Impact of independently controlling ventilation rate per person and ventilation rate per floor area on perceived air quality, sick building symptoms and decision making. Lawrence Berkeley National Laboratory; Allen, J., et al. (2015, submitted). The Combined and Independent Effects of Carbon Dioxide, Ventilation and Volatile Organic Compounds on the Cognitive Function of Office Workers in a Green Building."; Personal communications with Bill Fisk, Pawel Wargocki, and Joseph Allen.

34. NOAA's Pacific Marine Environmental Laboratory Carbon Dioxide Program. What is ocean acidification? Retrieved from www.pmel. noaa.gov; World Meteorological Organization. (2014, September 9) *WMO Greenhouse Gas Bulletin*; The Interacademy Panel. (2009, June). IAP statement on ocean acidification; National Oceanic and Atmospheric Administration. Coral reefs—an important part of our future. Retrieved from www.noaa.gov/features/ economic_0708/coralreefs.html; Veron, J. E. (2010, December 6). Is the end in sight for the world's coral reefs? *Yale environment360*; Oregon State University News & Research Communications. (2014, December 15) New study finds saturation state directly harmful to bivalve larvae.

35. Pimm, S., et al. (2014, May). The biodiversity of species and their rates of extinction, distribution, and protection. *Science*; Duke Environment News. (2014, May 29). New technologies making it easier to protect threatened species; Senckenberg Research Institute. (2011, August 24). Global warming may cause higher loss of biodiversity than previously thought, *ScienceDaily*.

36. Dell, M., et al. (2008, June). Climate change and economic growth: evidence from the last half century. National Bureau of

Economic Research; Solomon, S. (2009, February). Irreversible climate change due to carbon dioxide emissions. *Proceedings of the National Academies of Science*; Carty, T. (2012, September) Extreme weather, extreme prices. *Oxfam*.

37. U.S. Department of Defense. (2014). 2014 climate change adaptation roadmap; Femia, F., & and Werrell, C. (2012, February). Syria: climate change, drought and social unrest. The Center for Climate and Security; Cohen, J. (2015, March 2). A perfect storm: a UCSB scientist links a warming trend to record drought and later unrest in Syria, the U.C. Santa Barbara. *Current*; Holthaus, E. (2015, March 2). New study says climate change helped spark Syrian Civil War. *Slate*; National Oceanic and Atmospheric Administration News. (2011, October 27). NOAA study: human-caused climate change a major factor in more frequent Mediterranean droughts. Retrieved from www. noaanews.noaa.gov/stories2011/images/hoerlingetalfig1b.jpg; Warrick, J., & Pincus, W. (2008, September 10). Reduced dominance is predicted for U.S. *Washington Post*; Sample, I. (2009, March 18). World faces 'perfect storm' of problems by 2030, chief scientist to warn. *UK Guardian*.

38. Carrington, D. (2010, November 28). Climate change scientists warn of 4C global temperature rise. *The Guardian*; Purdue University news service. (2010, May 4). Researchers find future temperatures could exceed livable limits.

39. Hansen, J., et al. (2005, June). Earth's energy imbalance: confirmation and implications. *Science*.

40. Nordhaus, W. (1977, February). Economic growth and climate: the carbon dioxide problem. *The American Economic Review*; Hare, B., et al. (2014, October). Rebuttal of 'Ditch the 2°C warming goal'. Climate Analytics; Rahmstorf, S. (2014, October 1). Limiting global warming to 2°C—why Victor and Kennel are wrong. Retrieved from RealClimate.org; UN Framework

Convention on Climate Change. (2015, May 4). Report on the
structured expert dialogue on the 2013–2015 review. Retrieved
from unfccc.int/resource/docs/2015/sb/eng/inf01.pdf.

41. IEA. (2014, September). Capturing the Multiple Benefits of
Energy Efficiency; Wynn, G. (2009, November 10). Cost of
extra year's climate inaction $500 billion: IEA. *Reuters*; Gillis, J.
(2014, April 13). Climate efforts falling short, U.N. panel says.
New York Times.

42. Kanter, J., & Revkin, A. (2007, January 30). World scientists near
consensus on warming. *New York Times*; Gillis, J., & Chang, K.
(2014, May 12). Scientists warn of rising oceans from polar melt.
New York Times.

43. Carrington, D. (2014, November 26). Reflecting sunlight into space
has terrifying consequences, say scientists. *The Guardian*.

44. Elgie, S., & McClay, J. (2013, July 24). BC's carbon tax shift after five
years: results. Sustainable Prosperity, University of Ottawa.

45. Schmalensee, R., & Stavins, R. (2013, Winter). The SO2 allowance
trading system: the ironic history of a grand policy experiment.
Journal of Economic Perspectives; Shapiro, I., & Irons, J. (2011, April
12). Regulation, employment, and the economy. *Economic Policy
Institute*; U.S. EPA, Office of Air and Radiation. (2011, March). The
Benefits and Costs of the Clean Air Act from 1990 to 2020. Final
Report; "A Carbon Trading System Worth Saving." Editorial Board.
New York Times. 6 May 2013.

46. White House news release. (2014, November 11). U.S.-China Joint
Announcement on Climate Change; Podesta, J., & Holdren, J. (2014,
November 12). The U.S. and China just announced important new
actions to reduce carbon pollution. [White House blog] Retrieved
from www.whitehouse.gov/blog/2014/11/12/us-and-china-j
ust-announced-important-new-actions-reduce-carbon-pollution;
China sets cap on energy use. (2014, November 19). Retrieved
from ShanghaiDaily.com; China seeks to cap coal use at 4.2 billion

tonnes by 2020. (2014, November 19). *The Economic Times*; Chen, K. &, Stanway, D. (2014, November 18). China needs to cap coal use by 2020 to meet climate goals -think tank. *Reuters*; Puko, T., & Chuin-Wei, Y. (2015, February 26). Falling Chinese coal consumption and output undermine global market. *Wall Street Journal*.

47. Liptak, A. (2014, June 23). Justices uphold emission limits on big industry. *New York Times*.

48. Lean, G. (2010, April 23). General election 2010: Britain's silent, green revolution. *The Telegraph*; Carrington, D., & Goldenberg, S. (2009, December 4). Gordon Brown attacks 'flat-earth' climate change sceptics. *The Guardian*; Doyle, G., & Sumner, T. (2015, May 1). General election 2015: key points on energy policy. Retrieved from BLPlaw.com; Brownstein, R. (2010, October 9). GOP gives climate science a cold shoulder. *National Journal*; Davenport, C. (2015, March 19). McConnell urges states to help thwart Obama's 'War on Coal'. *New York Times*.

49. Gillis, J., & Schwartz, J. (2015, February 21). Deeper ties to corporate cash for doubtful climate researcher. *New York Times*.

50. Revkin, A. (2009, April 23). Industry ignored its scientists on climate. *New York Times*. (The 1995 draft primer retrieved from documents.nytimes.com/global-climate-coalition-aiam-climate-change-primer.); Greenpeace report. (2010, March 30). Koch industries: secretly funding the climate denial machine. Retrieved from www.greenpeace.org/usa/global-warming/climate-deniers/koch-industries; "Smithsonian Statement on Climate Change." (2014, October 2). Retrieved from newsdesk.si.edu/releases/smithsonian-statement-climate-change.

51. The Committee for Skeptical Inquiry. (2014, December 5). Deniers are not Skeptics. Retrieved from csicop.org; Davenport, C. (2014, November 10). Republicans vow to fight EPA and approve Keystone Pipeline. *New York Times*; Horsley, S. (2014, November

12). China, U.S. pledge to limit greenhouse gases. *NPR's Morning Edition*.

52. Fukushima disaster bill more than $105bn, double earlier estimate. (2014, August 27). Retrieved from RT.com/news; Wilson, W., et al. River Network Report. (2012, April). *Burning our rivers: the water footprint of electricity*.

53. Huntington, H. et al. Energy Modeling Forum (2013, September). Changing the game? Emissions and market implications of new natural gas supplies. *Stanford University*; Shearer, C., et al. (2014, September). The effect of natural gas supply on US renewable energy and CO2 emissions. *Environmental Research Letters*; Schneising, O., et al. (2014, October). Remote sensing of fugitive methane emissions from oil and gas production in North American tight geologic formations. *Earth's Future*.

54. Dale, M., & Benson, S. (2013, February). Energy balance of the global photovoltaic (PV) industry—Is the PV industry a net electricity producer? *Environmental Science & Technology*; Denholm, P., & Mehos, M. (2011, November). Enabling greater penetration of solar power via the use of CSP with thermal energy storage. *NREL*.

55. Global Wind Energy Council. (2015). Global statistics. Retrieved from Gwec.net; Carbajales-Dale, M., et al. (2014, February). Can we afford storage? A dynamic net energy analysis of renewable electricity generation supported by energy storage. *Energy & Environmental Science*.

56. The National Coal Council (2003, May). *Coal-Related Greenhouse Gas Management Issues*; Department of Energy. (2008). *Retrofitting the existing coal fleet with carbon capture technology*. As cited here: science.house.gov/sites/republicans.science.house.gov/files/documents/hearings/101311_Charter_0.pdf; Van Loon, J. (2014, December 4). This process averts climate change. Now the bad news. *BloombergBusiness*; (2010, November 9). Leaks from CO2 stored deep underground could contaminate drinking water.

Duke Environment; Elgin, B. (2008, June 18). The dirty truth about clean coal. *BloombergBusinessWeek;* Zobacka, M., & Gorelick, S. (2012, June). Earthquake triggering and large-scale geologic storage of carbon dioxide. *Proceedings of the National Academy of Sciences;* Slavin, T., & Jha, A. (2009, July 29). Not under our backyard, say Germans, in blow to CO2 plans. *The Guardian;* MIT CCS Technology Program (2015, January 5). Schwarze pumpe fact sheet: carbon dioxide capture and storage project. Retrieved from sequestration.mit.edu/tools/projects/vattenfall_oxyfuel.html.

57. Searchinger, T., & Heimlich, R. (2015, January). Avoiding bioenergy competition for food crops and land creating a sustainable food future. *World Resources Institute;* Lott, M. (2014, April 21). Corn-waste biofuels might be worse than gasoline in the short term. *Scientific American.*

58. "A Big Laser Runs Into Trouble" (Editorial). *New York Times.* 6 Oct 2012.

59. Replogle, M. A global high shift to public transport, walking, and cycling. Institute for Transportation and Development Policy. 17 September 2014; European Commission. Road transport: reducing CO2 emissions from vehicles. Updated May 2015.

60. Romm, J. (2006, November). The car and fuel of the future. *Energy Policy;* Flynn, P. (2002, June). Commercializing an alternate vehicle fuel: lessons learned from natural gas for vehicles. *Energy Policy;* Motavalli, J. (2006, July 30). P what? PZEV's are unsung heroes in the push to clean up the air. *New York Times.*

61. U.S. Department of Energy. (2014, September 15). The History of the Electric Car. Retrieved from energy.gov/articles/history-electric-car; Romm, J., & Frank, A. (2006, April). Hybrid vehicles gain tractionm. *Scientific American;* U.S. Department of Energy. (2014, May 30). Electric vehicle manufacturing taking off in the U.S; U.S. Department of Energy. Clean tech now. October 2013; Parkinson, G. (2014, August 21). Why EVs will make

solar viable without subsidies. Retrieved from Reneweconomy. com.au; Hummel, P., et al. (2014, August 20). Will solar, batteries and electric cars re-shape the electricity system? UBS report. Retrieved from knowledge.neri.org.nz/assets/uploads/ files/270ac-d1V0tO4LmKMZuB3.pdf; Lienert, P. (2015, March 26). Automakers rush to double electric car milage. *Christian Science Monitor*.

62. Romm, Joseph. *The Hype About Hydrogen*. Island Press, 2005. Print; Safety, Codes, and Standards Fact Sheet. U.S. DOE Office of Energy Efficiency and Renewable Energy, Fuel Cell Technologies Program. February 2011. Retrieved from energy.gov/sites/prod/ files/2014/03/f9/fct_h2_safety.pdf.

63. Ford Motor Company, Hydrogen Fuel Cell Vehicles. *Sustainability Report 2013/2014*; Ayre, J. (2014, November 19). Toyota to lose $100,000 on every hydrogen FCV sold? Retrieved from CleanTechnica.com; Courtenay, V. (2015, February 3). Hyundai cuts price of Tucson FCV 43% in South Korea. Retrieved from WardsAuto.com; Bringing Fuel Cell Vehicles to Market. California Fuel Cell Partnership Study. October 2001.

64. Gurwick, N., et al. (2012). The scientific basis for biochar as a climate change mitigation strategy: does it measure up? *Union of Concerned Scientists*; The U.S. Congressional Budget Office. The Potential for Carbon Sequestration in the United States. September 2007; Bailey, R., et al. Livestock: Climate Change's Forgotten Sector. Chatham House, The Royal Institute of International Affairs, Research Paper, December 2014.

65. Pope Francis. Letter Of His Holiness Pope Francis To The Prime Minister Of Australia On The Occasion Of The G20 Summit. The Vatican. November 2014.

66. Wilson, D. (2014, November 24). Special report: why metro Houston fears the next big storm. *Reuters*; Folger, T. (2013, September). Rising seas. *National Geographic*; Goodell, J. (2013, June 20).

Goodbye, Miami. *Rolling Stone*; McNamara, D., et al. (2015, March 25). Climate adaptation and policy-induced inflation of coastal property value. *PLoS One*.

67. Bailey, R., et al. (2014, December). Livestock—Climate Change's Forgotten Sector. Chatham House.
68. National Snow and Ice Data Center press release. (2014, October 7). Arctic sea ice continues low; Antarctic ice hits a new high.

INDEX